Hidden Persuaders in Cocoa and Chocolate

Hidden Persuaders in Cocoa and Chocolate

A Flavor Lexicon for Cocoa and Chocolate Sensory Professionals

Renata Januszewska

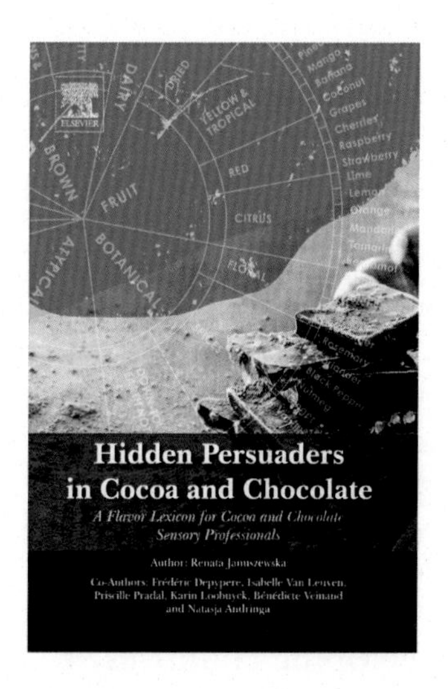

ELSEVIER

WP
WOODHEAD PUBLISHING
An imprint of Elsevier

Woodhead Publishing is an imprint of Elsevier
The Officers' Mess Business Centre, Royston Road, Duxford, CB22 4QH, United Kingdom
50 Hampshire Street, 5th Floor, Cambridge, MA 02139, United States
The Boulevard, Langford Lane, Kidlington, OX5 1GB, United Kingdom

Notices
Knowledge and best practice in this field are constantly changing. As new research and experience broaden our understanding, changes in research methods, professional practices, or medical treatment may become necessary.

Practitioners and researchers must always rely on their own experience and knowledge in evaluating and using any information, methods, compounds, or experiments described herein. In using such information or methods they should be mindful of their own safety and the safety of others, including parties for whom they have a professional responsibility.

To the fullest extent of the law, neither the Publisher nor the authors, contributors, or editors, assume any liability for any injury and/or damage to persons or property as a matter of products liability, negligence or otherwise, or from any use or operation of any methods, products, instructions, or ideas contained in the material herein.

Library of Congress Cataloging-in-Publication Data
A catalog record for this book is available from the Library of Congress

British Library Cataloguing-in-Publication Data
A catalogue record for this book is available from the British Library

ISBN: 978-0-12-815447-2 (print)
ISBN: 978-0-12-815448-9 (online)

For information on all Woodhead publications visit our website at
https://www.elsevier.com/books-and-journals

Publisher: Andre Gerharc Wolff
Acquisition Editor: Megan R. Ball
Editorial Project Manager: Susan E. Ikeda
Production Project Manager: Sojan P. Pazhayattil
Designer: Mark Rogers

Typeset by Thomson Digital

Contents

6. Botanical

7. Trigeminal Effects

8. Atypical

About Barry Callebaut Group

Zurich-based Barry Callebaut Group is the world's leading manufacturer of high-quality chocolate and cocoa products: from sourcing and processing cocoa beans to producing the finest chocolates, including chocolate fillings, decorations, and compounds. The Barry Callebaut Group runs close to 60 production facilities worldwide and employs a diverse and dedicated global workforce of about 11,000 people.

The Barry Callebaut Group serves the entire food industry, from industrial food manufacturers to artisanal and professional users of chocolate, such as chocolatiers, pastry chefs, bakers, hotels, restaurants, or caterers. The three global brands catering to the specific needs of these gourmet customers are Callebaut®, Cacao Barry®, and Carma®.

The Barry Callebaut Group is committed to make sustainable chocolate the norm by 2025, to help ensure future supplies of cocoa and improve farmer livelihoods. It supports the Cocoa Horizons Foundation in its goal to shape a sustainable cocoa and chocolate future.

Introduction

Chapter Outline

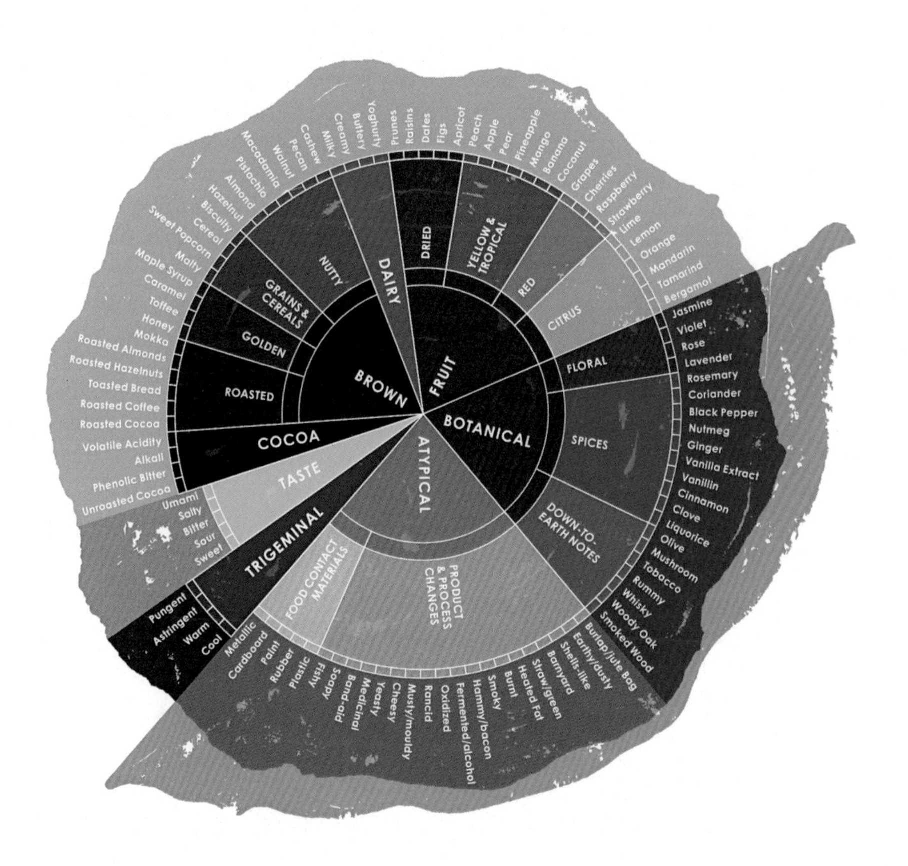

Hidden Persuaders in Cocoa and Chocolate. http://dx.doi.org/10.1016/B978-0-12-815447-2.00001-2

OBJECTIVE AND AIMS

The major objective of the book '*Hidden Persuaders in Cocoa & Chocolate*' is to popularize the knowledge about tastes and flavors describing cocoa and chocolate.

The aim is to enable different departments within industry to speak a common sensory language when describing cocoa and chocolate in a professional way – either for new product development or quality improvement. By using this book people can develop a higher awareness of tastes and flavors, and build up their overall sensory skills.

Another aim is to broaden the flavor language of cocoa and chocolate, by storytelling about ingredients in which tastes and flavors are found, and also bringing the tasting experience related to 'pairing flavors' based on on-line applications called 'Chocolizer™' and 'Itinero™'.

We are convinced that well-trained sensory panels are able to detect most of the cocoa and chocolate flavors. By using the extended knowledge documented in this book, the sensory leaders may broaden and deepen the skills of trained panels. Striving to add value to the cocoa and chocolate business, we believe that well trained panels are an excellent tool for guiding internal product development and improving communication with external business partners.

METHOD

The final selection and description of 105 sensory attributes presented in this book is based on a review of existing cocoa and chocolate flavor wheels and numerous scientific papers describing sensory evaluation of cocoa and chocolate.

The selected sensory attributes are grouped into eight flavor levels, which also correspond to the Cocoa and Chocolate Flavor Wheel:

1. *Taste* (sweet, sour, bitter, salty, umami),
2. *Cocoa* (unroasted cocoa, alkali, phenolic bitter, volatile acidity),
3. *Brown* (four sub-categories: roasted, golden, grains and cereals, nutty),
4. *Dairy* (milky, creamy, buttery, yoghurty),
5. *Fruit* (four sub-categories: dried, yellow and tropical, red, citrus),
6. *Botanical* (three sub-categories: floral, spices, down-to-earth),
7. *Trigeminal* (cool, warm, astringent, pungent),
8. *Atypical* (two sub-categories: product and process changes, food contact materials).

This book describes eight flavor levels of sensory perception and none of these levels is considered as more or less important, because all of them contribute to a unique perception of sensory profiles in cocoa and chocolate products. This flavor approach does not include the following aspects of chocolate evaluation: appearance, sound, touch, and texture.

For each of the eight flavor levels, some or all of following ten descriptive parts are developed:

Description	Flavor Levels							
	Taste	Cocoa	Brown	Dairy	Fruit	Botani-cal	Tri-geminal	Atypical
1. Sensory definition	■	■	■	■	■	■	■	■
2. Key taste compound[1]	■	■						
3. Taste description	■	■						
4. Key aroma compound(s)[2]		■	■	■	■	■	■	■
5. Flavor description		■	■	■	■	■		
6. Notes description						■		
7. Trigeminal effect							■	
8. Smell description		■	■	■	■	■		■
9. Chocolate tasting[3]	■	■	■	■	■	■	■	■
10. Emotions[4]	■	■	■	■	■	■	■	■

[1]The most important compound contributing to a specific taste.
[2]The most important volatile compound(s) contributing to a specific aroma.
[3]Stories about chocolate and/or specific flavor ingredient and suggestions for pairing with chocolate. *http://www.callebaut.com/benl/chocolizer; http://www.cacao-barry.com/itinero*
[4]Positive/negative emotions perceived during tasting a specific flavor.

Every author of the book contributed to a unique part of the presented information. Barry Callebaut's experts are primary responsible for cocoa and chocolate specific information regarding sensory, marketing, and chemical composition of each flavor. Please note that we have selected only a few compounds per each taste and flavor, as well as off-flavor, acknowledging the whole complex mixture of chemical compounds behind each of them. Per selected chemical compound, structural formula, and molecular formula are reported in the book.

While positive emotions, linked to tastes, flavors, and notes, come from extensive research of Givaudan's experts, negative emotions are linked to off-flavors researched by Barry Callebaut's scientists, during multiple shelf-life studies and testing for new product development.

DEFINITIONS: TASTE – ODOR/SMELL – AROMA – FLAVOR – NOTES/NUANCES/HINTS

Taste, being perceived on the tongue, can be expressed as 'sweet', 'bitter', 'sour', 'salty' or 'umami'.

Odor (or smell) is perceived explicitly through the nose, while aroma is a sensory perception via the nose and mouth together, so called 'retronasal' perception. As a consequence, both taste and aroma are a part of 'flavor'.

A full flavor perception consists of taste, smell, and aroma, as well as special perception – often referred to as 'chemical feeling factors'—that happens through triggering of a trigeminal nerve. This nerve reacts to pain and change in temperature, being responsible for sensations such as 'cool', 'warm', 'astringent' or 'pungent'. These trigeminal factors contribute to 'mouthfeel perception' that also includes all kinds of structure and texture related impressions, which are not described in this material.

By notes, nuances or hints we understand a threshold-level perception of a certain sensory attribute. For instance, most volatile compounds, called 'potent aroma compounds' are present in chocolate in very low levels, and are detectable by highly trained panelists or the most sensitive individuals, only. These impressions are analogous to perfumers' 'top notes'.

HOW TO USE THE BOOK?

With this book we aim to enrich and further develop the sensory language of cocoa and chocolate. The content can be used during various cocoa and chocolate sensory trainings.

The internal trainings may be organized by sensory specialists in Quality Assurance, who continuously strive to improve the performance of sensory panels. The external trainings can be prepared for customers or consumers, by product developers in R&D, chefs in Chocolate Academy centers, and other employees, who want to study and practice flavor evaluation in a professional way.

While purposely tasting chocolate via the so called 'Chocolate Tasting Ritual', all five human senses are involved. Firstly, *sight* is crucial. Chocolates may surprise by their unique color, perfect gloss or appealing shape. *Touch* is the second sense that matters: attributes like hardness and finger sensitivity contribute to the sensory experience. Closely associated to hardness is the third sense: *sound*. Snap is the unique feature to describe chocolate's sound upon breaking. *Smell*, or the direct perception of volatile components by the nose, is a powerful way to differentiate. Both smell intensity and smell complexity (amount of aromas that can be perceived together) play a role. And last but not least, the sense of *taste* is experienced when taste and aroma molecules are being released during oral processing of chocolate. While this book covers the most relevant attributes linked to two out of five senses of this Chocolate Tasting Ritual (smell and taste), it is acknowledged that all senses together shape the sensory perception of a particular cocoa or chocolate product. To implement such Chocolate Tasting Ritual one can consider some *Barry Callebaut (BC) chocolates* mentioned throughout this book.

The primary source of information used in this book consists of R&D and QA experience and knowledge. The secondary sources include existing literature about sensory evaluation of cocoa and chocolate. At the end of each chapter's introduction, the articles and other references are indicated.

We thank all Barry Callebaut and Givaudan colleagues who shared their critical comments and helped in developing this material:

Alex Landuyt (R&D, BC Belgium), Anais Maffard (Sourcing, BC Switzerland), Bob Van Overklift (student, BC Belgium), Clara Cabus (La Morella Nuts, Spain), Claudia Radi (R&D, BC Germany), Evelyne Maes (R&D, BC Belgium), Hanspeter Haefliger (Sourcing, BC Switzerland), Ive De Ruysscher (R&D, BC Belgium), Jan Schoen (Givaudan, Switzerland), Jemilatu Boye-Okit (QA, BC Belgium), Karen Maes (Marketing, BC Belgium), Marijke De Brouwer (R&D, BC Belgium), Nadine De Latthauwer (QA, BC Belgium), Philippe Troplin (R&D, BC France), Qiannan Cao (QA, BC China), Rosa Gimeno (La Morella Nuts, Spain), Serge Van

Berg (R&D, BC Belgium), Stephanie Garciaherreros (MK, BC Belgium), Thierry Bronchart (Gourmet, BC Belgium), Timothy Holman (Legal, BC Switzerland).

The following external reviewers are cordially thanked for opinions, suggestions and corrections:

Prof. Dr. Carolina Chaya – University of Barcelona, Spain

Prof. Dr. Dorota Majchrzak – University of Vienna, Austria

Prof. Dr. Hal MacFie – University of Reading, UK

Prof. Dr. Pascal Schlich – research director at INRA, University of Bourgogne, France, France

Prof. Dr. Jacques Viaene – University of Gent, Belgium

Dr. Jordi Ballester-Perez – University Institute of Vine and Wine, University of Bourgogne, France

Dr. Boris Gadzov – Sensory director, FlavorActiV Ltd, UK

Dr. Howard Moskowitz – President of Moskowitz Jacobs Inc., NY, USA

Jagoda Mazur – Senior sensory scientist, McCain foods Ltd., Toronto, Canada

Hilde Vanaerde – Sensory expert, VG Sensory, Belgium

Lieve Claes – Sensory expert, Haystack, Belgium

We wish All Readers of this book a pleasant time and efficient use of this material.

Author

Renata Januszewska, *PhD, Global R&D Sensory Methodologies Manager, Barry Callebaut*

Coauthors

Frédéric Depypere, *PhD, Global R&D Program Head & Global Sensory Lead, Barry Callebaut*

Isabelle Van Leuven, *PhD, GC-MS Specialist, Barry Callebaut*

Karin Loobuyck, *Consumer & Market Insights Manager EMEA, Barry Callebaut*

Priscille Pradal, *QA Sensory Evaluation Manager, Barry Callebaut*

Natasja Andringa, *Senior Food Technologist, Givaudan Int. SA*

Bénédicte Veinand, *Consumer Sensory Insights EAME, Sweet Goods Manager, Givaudan Int. SA*

Chapter 1

Taste

Chapter Outline

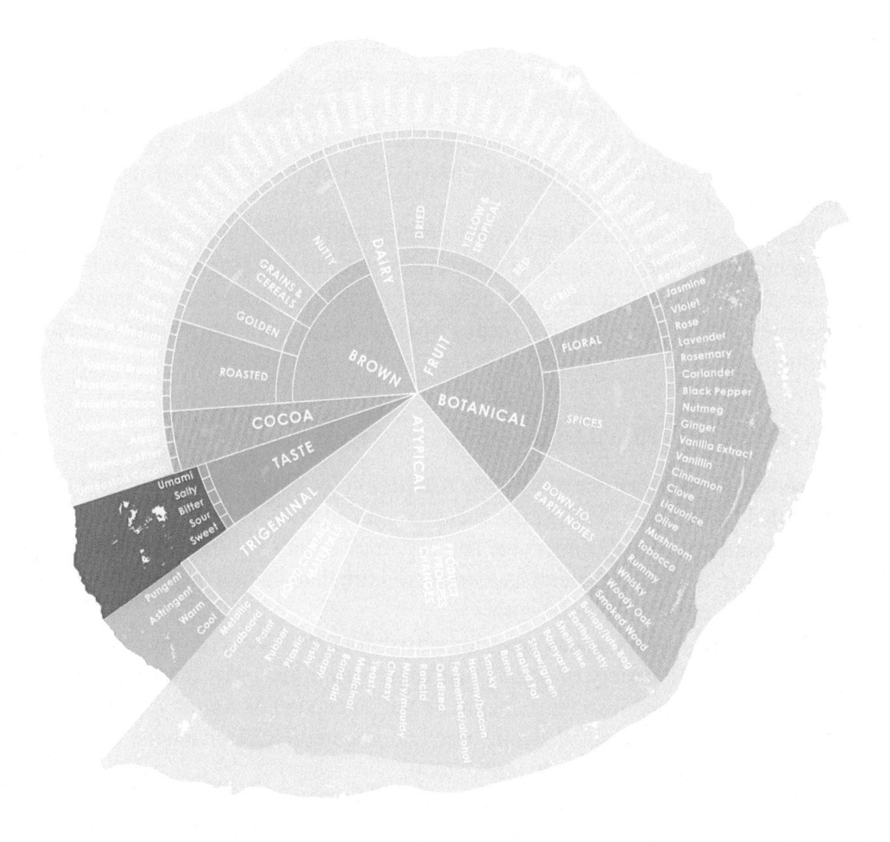

Hidden Persuaders in Cocoa and Chocolate. http://dx.doi.org/10.1016/B978-0-12-815447-2.00002-4

HISTORICAL PERSPECTIVE OF 'FIVE BASIC' TASTES

The subject of taste has intrigued scientists, philosophers, and epicureans for centuries. Historically, taste was defined as a philosophical question.

Ancient Greek philosophers like Plato perceived six tastes – bitter, sweet, salty, sour, astringent, and pungent. Aristotle made the matter much more complex by mapping taste between two poles. He placed sweet and bitter on opposite ends of the spectrum with salty, astringent, pungent, harsh, and acidic somewhere in between. Finally, Theophrastus, who studied under both philosophers added oily to the mix making in total eight basic tastes.

During the 18th century, the Enlightenment was in full force in Europe, and scientists tried to challenge many existing, common beliefs. Taste remained incredibly difficult to define. Linnaeus named 11 basic tastes (bitter, fatty, sour, astringent, sweet, salty, sharp, viscous, insipid, aqueous, and nauseous) while Polycarpe Poncelet equated different tastes to the art of harmonizing music notes of bitter, acid, sweet, peppery, astringent, sweet and sour, and weak or tasteless.

The 19th century reinforced only four tastes. Simplicity was preferred, and Fick narrowed the tastes down to sweet, salty, bitter, and sour. In addition, taste buds were discovered when tongue cells were examined under a microscope. Taste buds looked like keyholes that bits of food might fit into. We may conclude that we had known about the four basic tastes (sweet, sour, salty, and bitter) for centuries.

Umami – as the candidate for the official fifth taste – only came onto the scene in the twentieth century. Taste scientists were debating over validity of umami for decades because every new taste must fulfil strict criteria, including a unique quality and exclusive means of transduction, known as the receptor mechanism.

TASTE RECEPTOR MECHANISM

The sense of taste is mediated by *taste receptor cells*, which are bundled in clusters called 'taste buds'. These taste receptor cells identify oral concentrations of a large number of small molecules and report a sensation of taste to centres in the brain. Up to date the scientists identified more than 40 human TAS2R genes that function as *bitter* taste receptors.

TAS1R2+TAS1R3 genes play an active role in *sweetness* perception by binding to a wide variety of sugars and sugar substitutes. The mechanisms for *saltiness* (via epithelial sodium channel/ENaC receptor) or *sourness* perception (hyperpolarization-activated cyclic nucleotide-gated/HCN channels: ACCN1 and TASK-1) are still not fully understood.

The discovery of specific TAS1R1 and TAS1R3 protein receptors solved a long lasting scientific dilemma of the *umami* taste. This protein located in taste buds binds to the prototypical umami tastant and initiates the nerve signal to the brain that we interpret as umami.

During recent years, other candidates of primary taste – such as *fat* and *kokumi* are being discussed. The evidence for fat, as a primary taste, is reasonably strong. A possible CD36 taste receptor for fat has been already identified. Sensory scientists and nutritionists showed almost 20 years ago that we respond to fat in the mouth, with a rise of fats (triglycerides) in the blood. In case of kokumi, defined as a quality of continuity, mouth fullness and thickness, the evidence is weak and requires further research.

ABBREVIATIONS

TAS1R (#) = type 1, taste receptor, sweet
TAS2R = type 2, taste receptor, bitter
ENaC (epithelial sodium channel) receptor = taste receptor, salty
CD36 (cluster of differentiation 36) receptor = taste receptor, fatty
Hyperpolarization-activated cyclic nucleotide-gated (HCN) channels:
ACCN1 (amiloride-sensitive cation channel 1) = acid-sensing ion channel = taste receptor, sour
TASK-1 = acid-sensitive K+ (potassium) channel = taste receptor, sour

Sweet

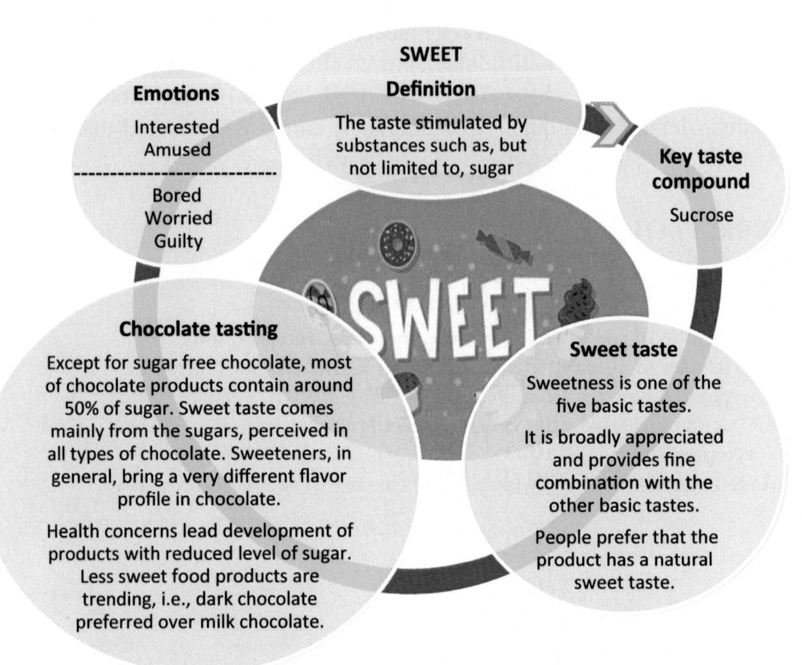

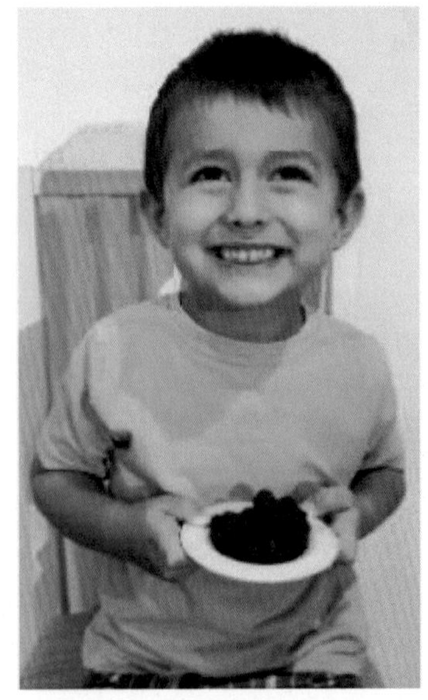

Innate reaction to sweetness: liking, happiness and relax.

Sucrose ($C_{12}H_{22}O_{11}$)

Sour

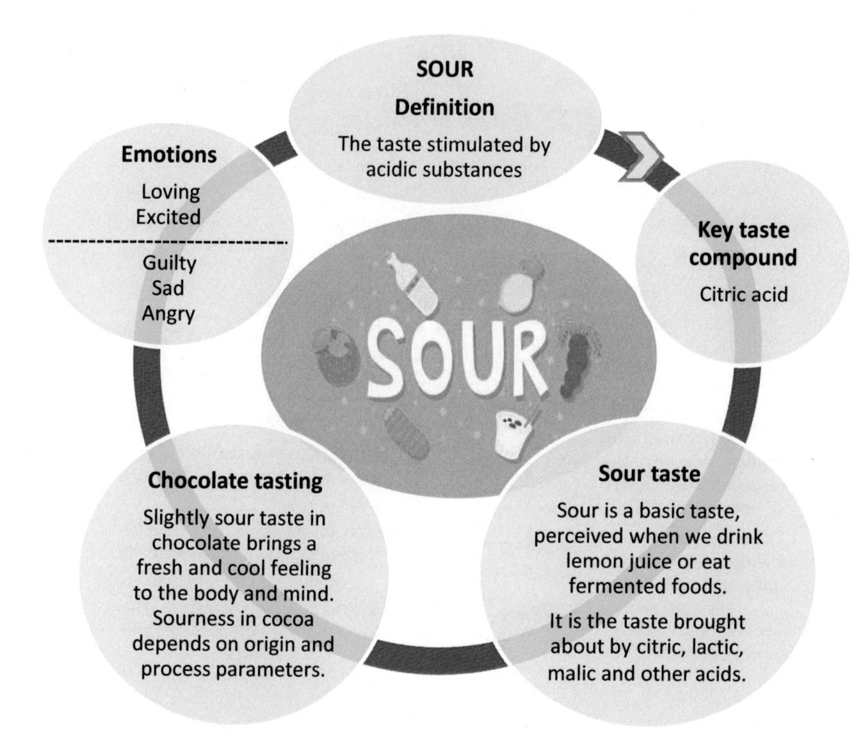

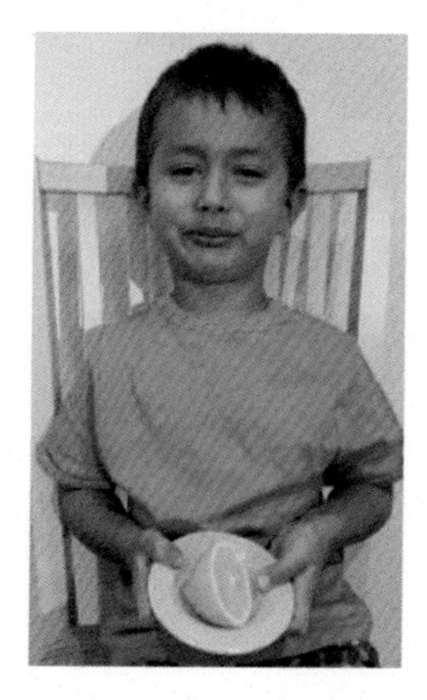

Innate reaction to sourness: disliking, aversion and stress

Citric acid ($C_6H_8O_7$)

Bitter

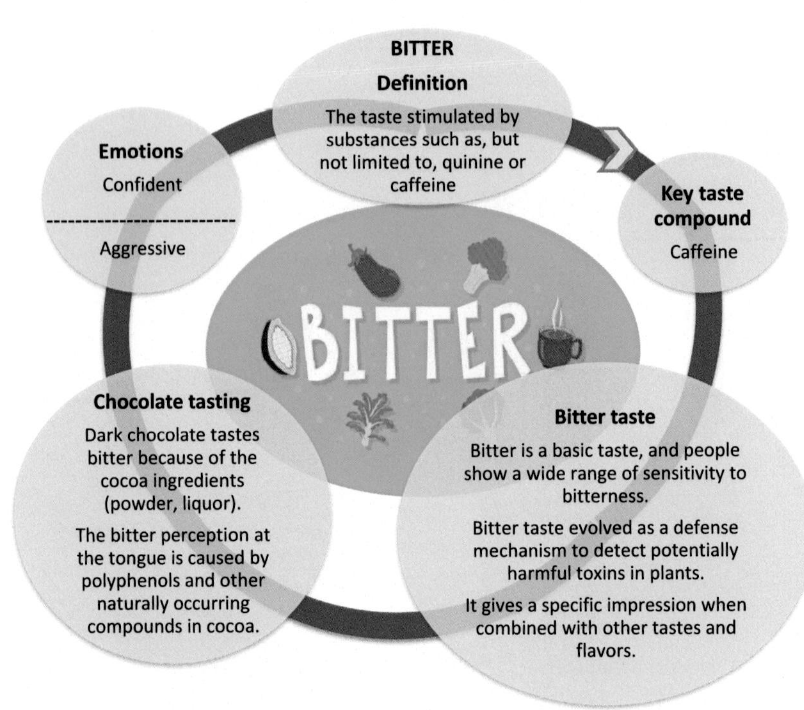

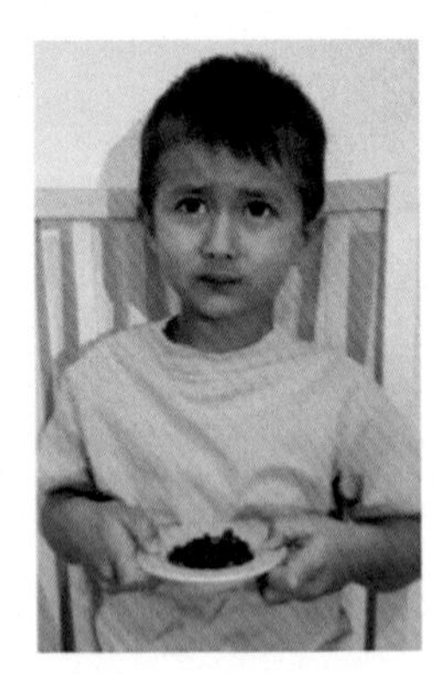

Innate reaction to bitterness: disliking, anger and aggression.

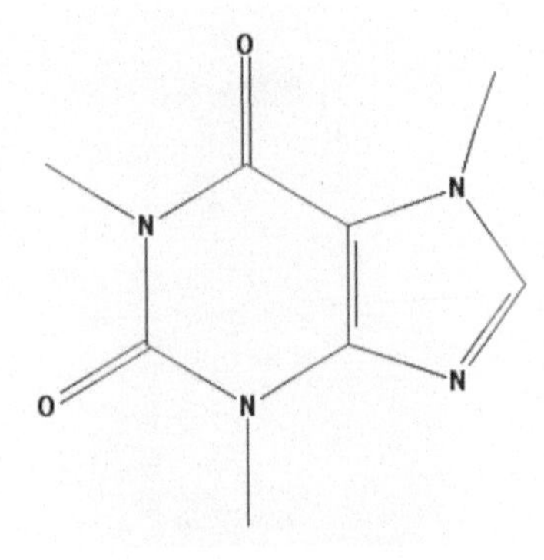

Caffeine ($C_8H_{10}N_4O_2$)

Salty

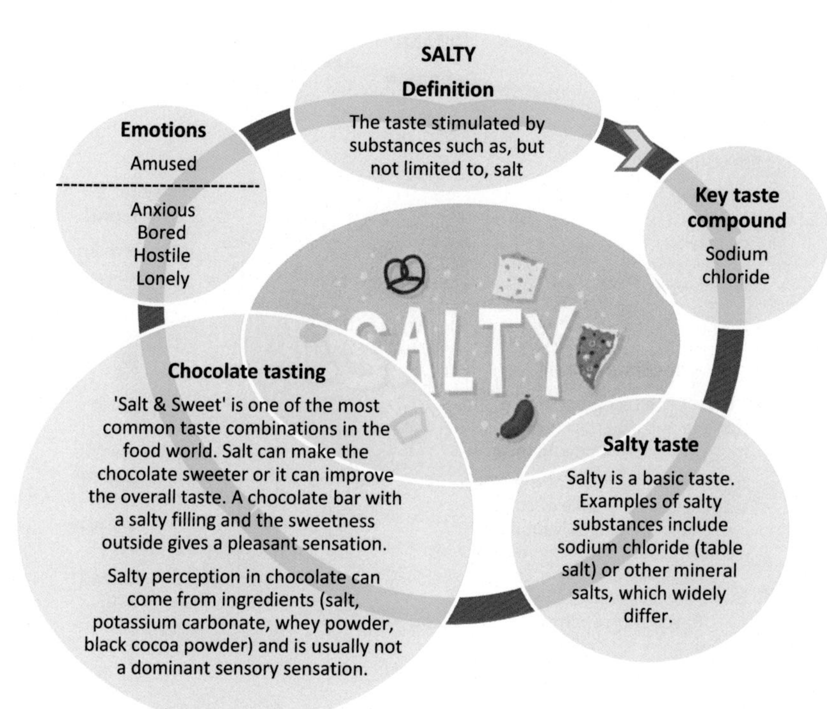

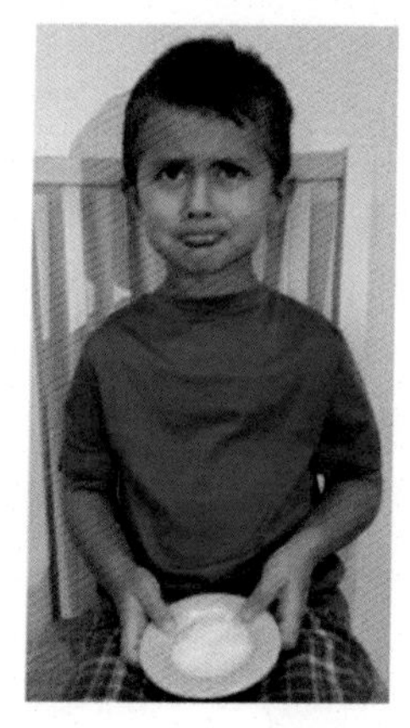

Innate reaction to saltiness: amusement and interest.

Na^{+} Cl^{-}

Sodium chloride (NaCl)

Umami

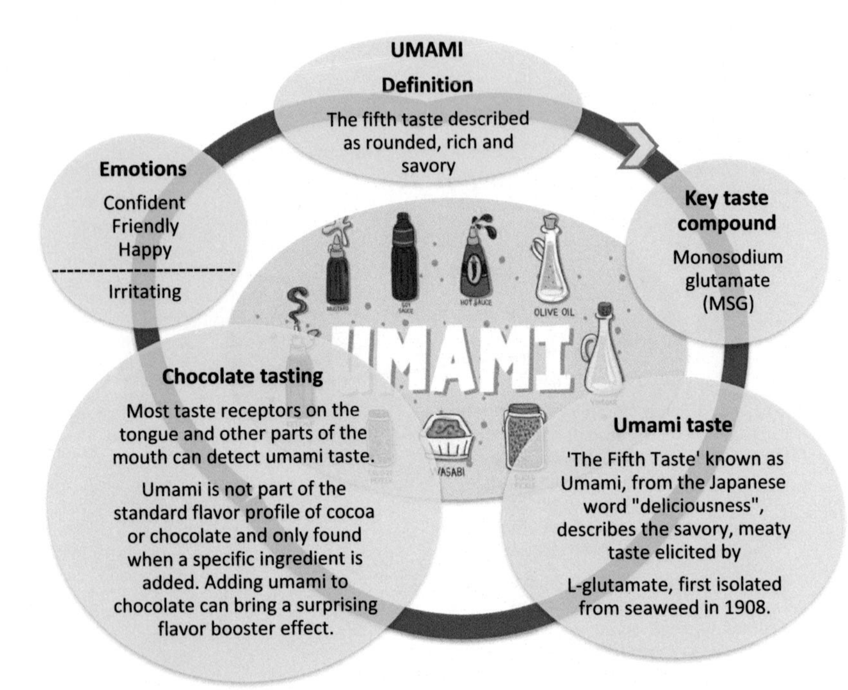

Innate reaction to umami: liking, interest and relax.

Monosodium glutamate ($C_5H_8NO_4Na$)

FURTHER READING

Afoakwa, E.O., 2010. Chocolate Science and Technology. Wiley-Blackwell, Oxford, (receptors).

Aprotosoaie, A.C., Luca, S.V., Miron, A., 2016. Flavor chemistry of cocoa and cocoa products—an overview. Compr. Rev. Food Sci. Food Saf. 15, 73–91, (chemical compounds).

Berry, D. (2014). Removing sugar while keeping the sweetness. Food Business News: Nov. 12 (sweet).

Burdock, G.A., 2010. Fenaroli's Handbook of Flavor Ingredients, sixth ed. CRC Press. Taylor & Francis Group, Boca Raton, NY, pp. 1-2162 (receptors).

Dalton, P., Maute, C., Oshida, A., Hikichi, S., Izumi, Y., 2008. The use of semantic differential scaling to define the multidimensional representation of odors. J. Sens. Stud. 23, 485–497, (smell of emotions).

Drake, M.A., Civille, G.V., 2002. Flavor lexicons. Compr. Rev. Food Sci. Food Saf. 2, 33–40, (receptors).

Gilbert, A.N., Fridlund, A.J., Lucchina, L.A., 2016. The color of emotion: A metric for implicit color associations. Food Qual. Prefer. 52, 203–210, (color of emotions).

Green, B.G., Alvarez-reeves, M., George, P., Akirav, C., 2005. Chemesthesis and taste: Evidence of independent processing of sensation intensity. Physiol. Behav. 86, 526–537, (receptors).

Hayes, J.E., Feeney, E.L., Allen, A.L., 2013. Do polymorphisms in chemosensory genes matter for human ingestive behavior? Food Qual. Prefer. 30, 202–216, (sweet).

Höhl, K., Schönberger, G.U., Busch-Stockfisch, M., 2014. Stimulus and recognition thresholds for the basic tastes in deionized water. Sci. Res. 3, 130–136, (sweet).

Jacob, N., Golmard, J.-L., Berlin, I., 2014. Differential perception of caffeine bitter taste depending on smoking status. Chemosens. Percept. 7 (2), 47–55, (sweet, sour, salty, bitter).

Jager, S., Schlich, P., Tijssen, I., Yao, J., Visalli, M., de Graaf, C., Stieger, M., 2014. Temporal dominance of emotions: measuring dynamics of food-related emotions during consumption. Food Qual. Prefer. 37, 87–99, (emotions of specific flavors/notes).

Kaneko, S., Kumazawa, K., Masuda, H., Henze, A., Hofmann, T., 2006. Molecular and sensory studies on the umami taste of Japanese green tea. J. Agric. Food Chem. 54 (7), 2688–2694, (umami).

Sandell, M.A., Breslin, P.A.S., 2006. Variability in a taste-receptor gene determines whether we taste toxins in food. Curr. Biol. 16, R792–R794, (taste receptors, bitter).

Spinelli, S., Masi, C., Dinnella, C., Zoboli, G.P., Monteleone, E., 2014. How does it make you feel? A new approach to measuring emotions in food product experience. Food Qual. Prefer. 37, 109–122, (classification of positive vs. negative emotions).

Wang, Q.J., Wang, S., Spenc, S., 2016. Turn up the taste': assessing the role of taste intensity and emotion in mediating crossmodal correspondences between basic tastes and pitch. Chem. Senses Vol. 41, 345–356, (tastes vs. emotions).

Chapter 2

Cocoa

Chapter Outline

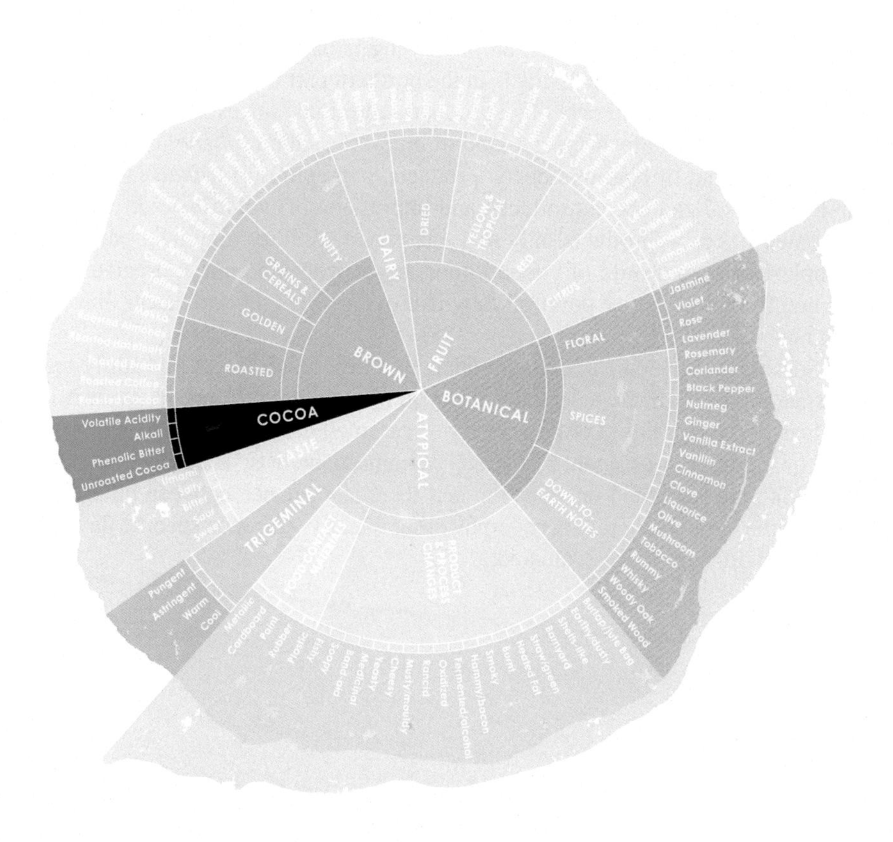

Hidden Persuaders in Cocoa and Chocolate. http://dx.doi.org/10.1016/B978-0-12-815447-2.00003-6

Perception of '*unroasted cocoa*' flavor is influenced by multiple factors: origin of cocoa beans, botanical variety of the beans, farmers agricultural practices, etc. All of these influencers are at tropical level and need different experts at every level of product growing and processing.

In the frame of sustainability, fair trade, bio-production schemes, and other important approaches, there is a growing awareness about the necessity of quality control and quality assurance at the cocoa production level.

Once beans are harvested, the initial processing at the farmer level starts, including fermentation and drying. Hereby, the farmers' knowledge and skills are key to create the expected quality of cocoa. Without a proper fermentation, the precursors of the cocoa flavor have no chance to be developed. Fermentation is therefore considered as a crucial step in reaching the right cocoa profile. Also the subsequent drying process needs to happen in a right way, otherwise some off-flavors may be formed (please refer to the last chapter 'Atypical flavors').

After the beans are dried, they are ready for transportation to the collection points, and then they are shortly stored and further transported overseas. It is interesting to notice that the initial processing is happening in the equatorial region while the main processing is in the northern part of Earth.

In case of cocoa and chocolate products, additional tastes are important to mention. They include: a special kind of bitterness called '*phenolic*' – typical for high content of phenolic cocoa products; perception of '*alkali*' – associated with alkalized cocoa products; and impression of '*volatile acidity*' – appreciated in cocoa products only to some extent. Acidity may be a part of a complex sensory profile of cocoa liquor, but it would be not appreciated or wanted from the quality point of view if it overpowers other pleasant flavors and notes.

Sensory evaluation of cocoa products requires defining and uniformly understanding each of the selected sensory attributes. When a cocoa product is evaluated in either liquid or powder form; the smell is a first key aspect of sensory quantitative assessment by a group of trained panelists. The taste is usually following the smell and the question is if there is a balance between bitterness and sourness, or one of them is overpowering. Finally, all cocoa specific flavors, notes and off-flavors are evaluated.

UNROASTED COCOA

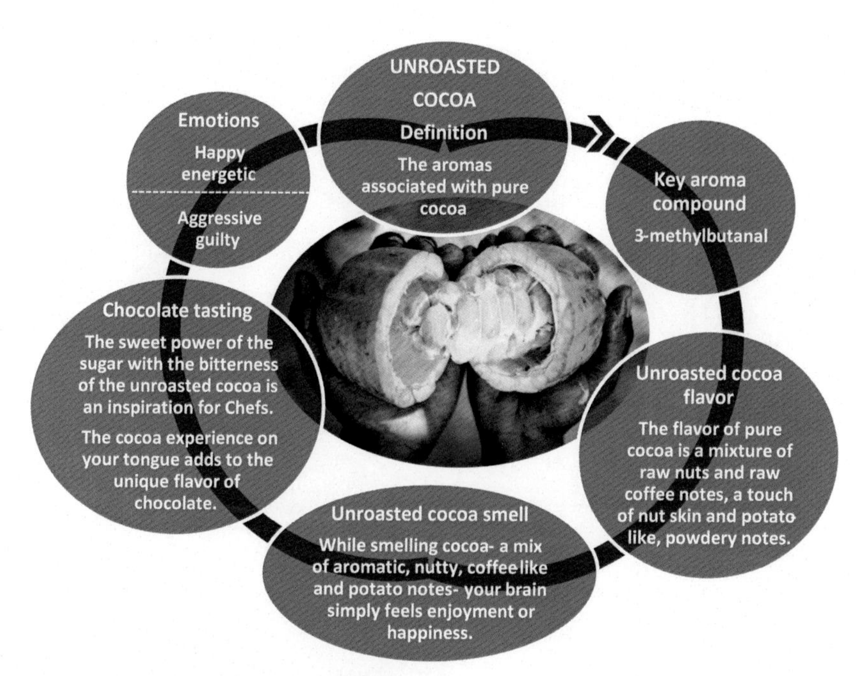

Cocoa trees with ripe fruits

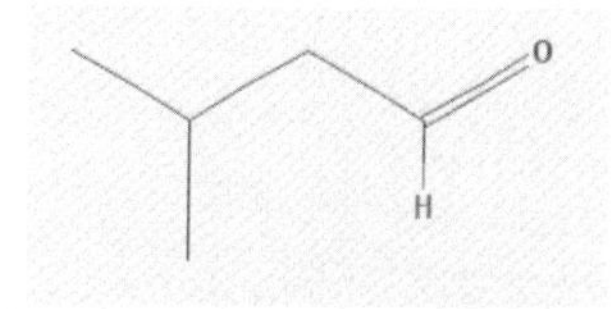

3-methylbutanal ($C_5H_{10}O$)

PHENOLIC BITTER

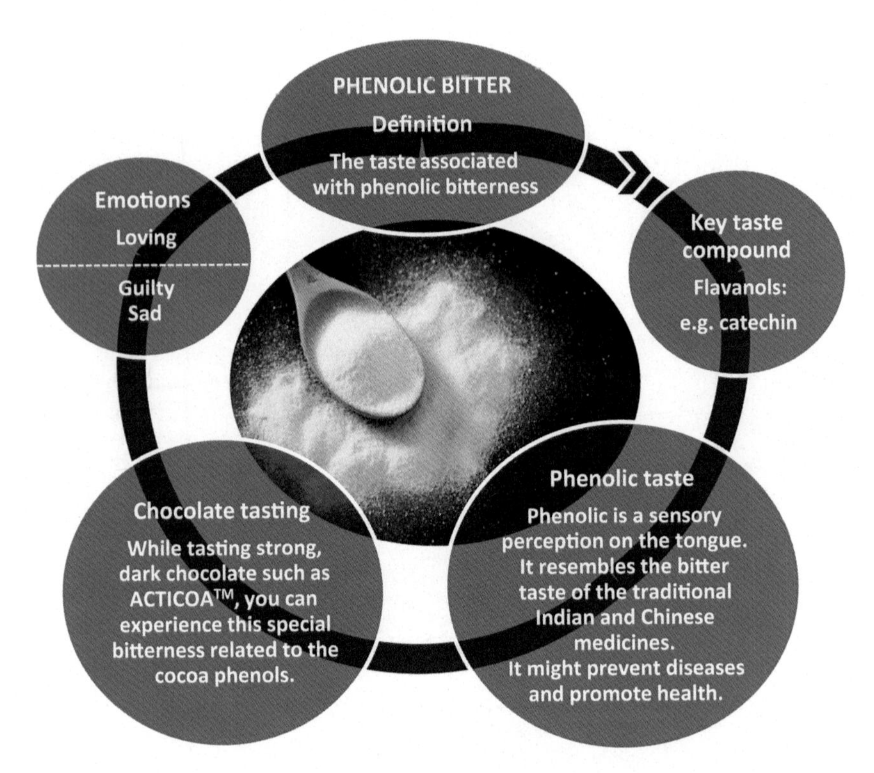

Flavanols: e.g. catechin ($C_{15}H_{14}O_6$)

ALKALI

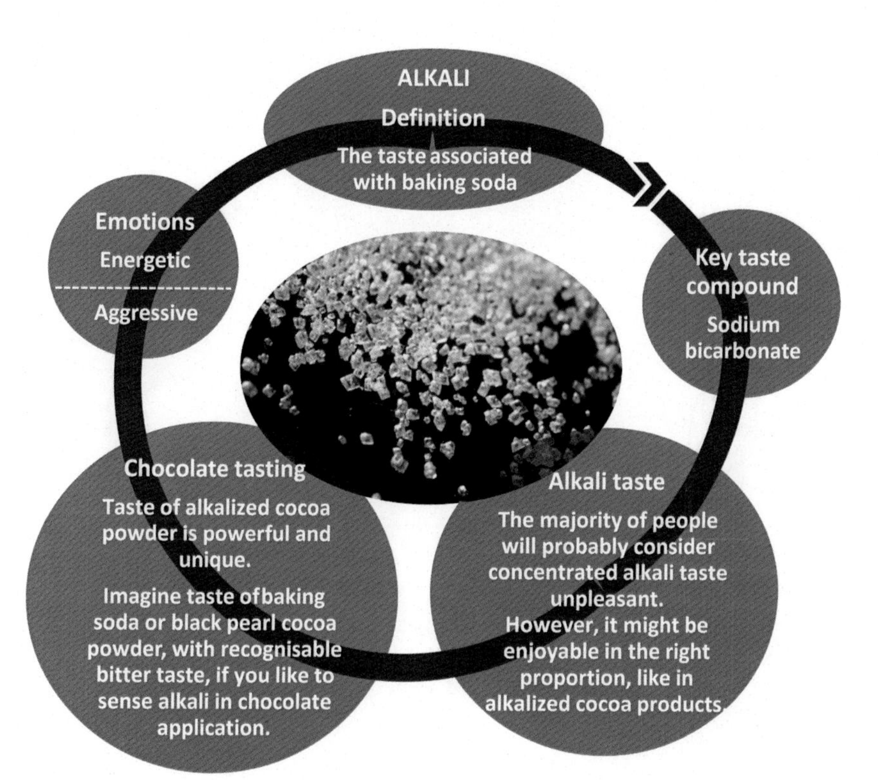

Sodium bicarbonate ($CHNaO_3$)

VOLATILE ACIDITY

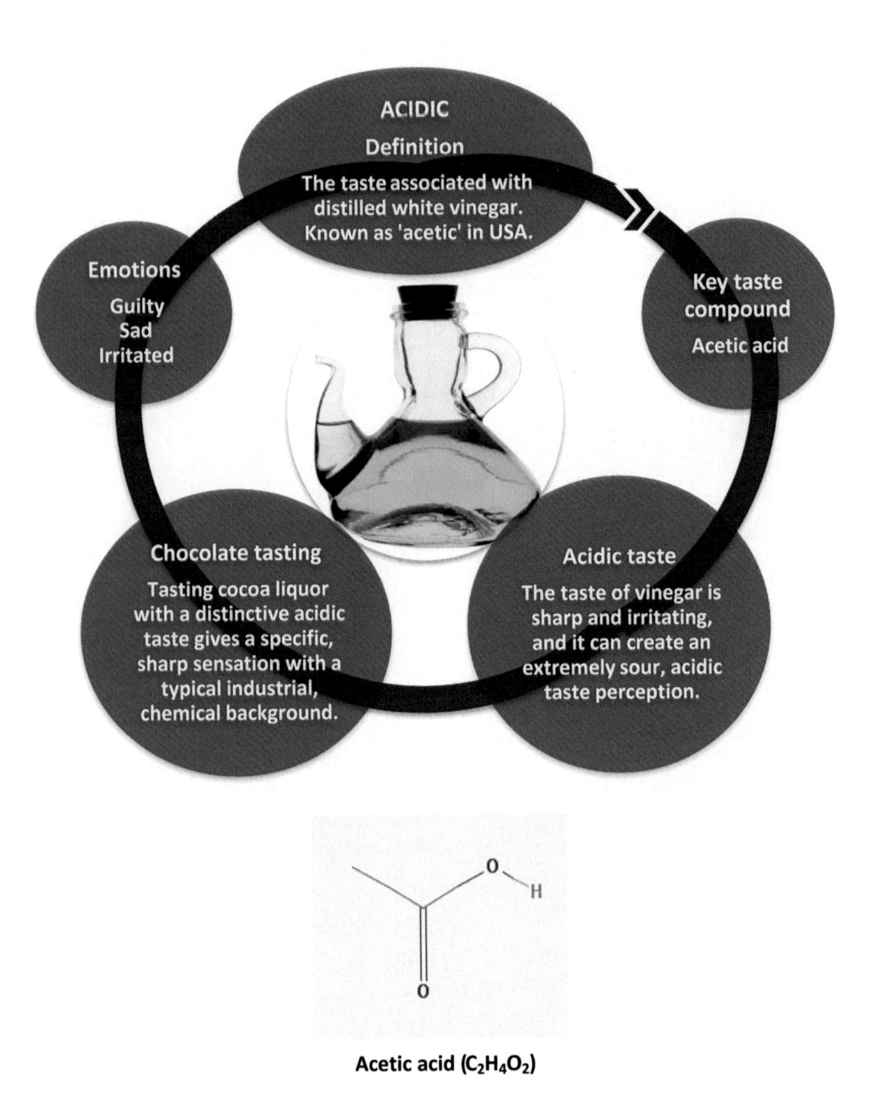

Acetic acid ($C_2H_4O_2$)

FURTHER READING

Jinap, S., Dimick, P.S., Hollender, R., 1995. Flavor evaluation of chocolate formulated from cocoa beans from different countries. Food Control 6 (2), 105–110, (cocoa).

Kongor, J.E., Hinneh, M., Van de Walle, D., Afoakwa, E.O., Boeckx, P., Dewettinck, K., 2016. Factors influencing quality variation in cocoa (*Theobroma cacao*) bean flavor profile: a review. Food Res. Int. 82, 44–52, (cocoa).

Reed, S., 2010. Sensory Analysis of Chocolate Liquor. Wilbur. An American Original. Cargill, Incorporated, Minnesota, United States, (cocoa).

Thorngate, J.H., Noble, A.C., 2006. Sensory evaluation of bitterness and astringency of 3R(-)-epicatechin and 3S(+)-catechin. J. Sci. Food Agric. 67 (4), 531–535, (phenolic).

Tran, P.D., Van de Walle, D., De Clercq, N., De Winne, A., Kadow, D., Lieberei, R., Messens, K., Tran, D.N., Dewettinck, K., Van Durme, J., 2015. Assessing cocoa aroma quality by multiple analytical approaches. Food Res. Int. 77, 657–669, (acidic).

Chapter 3

Brown

Chapter Outline

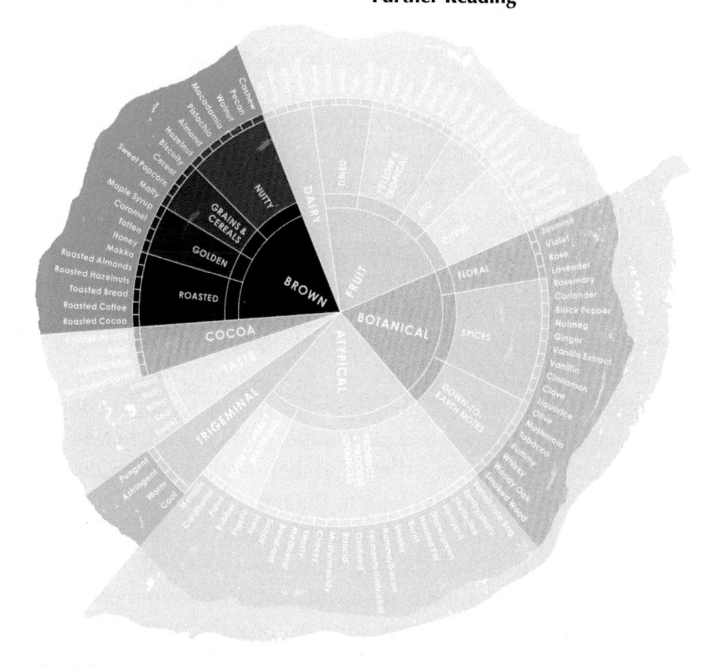

Hidden Persuaders in Cocoa and Chocolate. http://dx.doi.org/10.1016/B978-0-12-815447-2.00004-8

Once beans arrive to Europe, North America, or Asia they are usually stored in the warehouses. Appropriate storage conditions are again influencing the quality of the beans. Further, the beans are processed with the following steps: selection, blending, roasting, winnowing (removal of shells), grinding the cocoa nibs, obtaining cocoa mass, further pressing of cocoa mass to obtain cocoa powder and cocoa butter.

After the cocoa products (butter, powder, and mass, also called liquor) are prepared, other ingredients are added. Addition of these ingredients depends on the type of chocolate to be produced. A dark chocolate will include sugar and emulsifier, such as lecithin. A milk chocolate will have additionally a milk powder. A white chocolate will have a milk powder and sugar but no cocoa powder or cocoa liquor; as it contains only cocoa butter.

Looking from the sensory evaluation point of view, it is well known that there are at least two key steps influencing formation of flavors in this whole process of chocolate making.

The first step is roasting of cocoa beans. During roasting, a chemical process called 'Maillard reaction' starts. It is a very complex process during which sugars and amino acids react at higher temperature resulting in a typical '*roasted cocoa*' flavor. This flavor is mainly defined by the presence of pyrazines, pyridines and pyrroles. Since the same reaction happens when *hazelnuts, almonds, coffee, bread, mokka* are subjected to a high temperature, the roasted products have common compounds, which we call 'Roasted'.

The second step is conching, that is, mixing of all ingredients in big kneading and shearing machines where all ingredients are step-wise added and the whole process is carefully monitored regarding time and temperature. During this process, the level of volatile acids decreases and a more balanced final aroma develops. If the conching process is too long or too high temperatures are applied, most of the positive, volatile compounds may be lost. In this sense, the producers take care when manufacturing special cocoa origin chocolate, where, for example, fruity and floral notes are especially precious and desired.

Naturally, certain aromas in chocolate are linked to certain ingredients. Each ingredient gives specific aromatics often referred to as notes, nuances and perceptions. These perceptions are never dominant in the chocolate and will be detected by highly trained sensory panelists.

'Golden' aromas may be perceived as: *honey, caramel, toffee, maple*. 'Grains and Cereals' aromas include: *cereal, biscuity, sweet popcorn, malty*. Finally, 'Nutty' notes can be found from the following ingredients: *hazelnut, almond, walnut, pistachio, pecan, macadamia, cashew, coconut*. In this chapter, special flavors and notes are described and their taste in chocolate is explained.

The above flavors certainly belong to the most appreciated ones by the consumers. From numerous research projects focusing on association of sensory perceptions and emotions, we can see that consumers link these flavors with love, happiness, excitement, relaxation, interest, inspiration, cheerfulness, satisfaction, or even perception of some kind of 'beautifulness'.

Sensory evaluation of chocolate requires defining and uniformly understanding each of the selected sensory attributes. Evaluation of the smell can be a first part of sensory profiling, however the major focus is on taste, flavors, notes, and mouthfeel. Hereby, sweetness, sometimes saltiness and umami, can be added to the assessment list. A complete profiling test includes 20 up to 40 carefully selected sensory descriptors depending on the product type; most of them explained in the following chapters.

ROASTED

Roasted cocoa

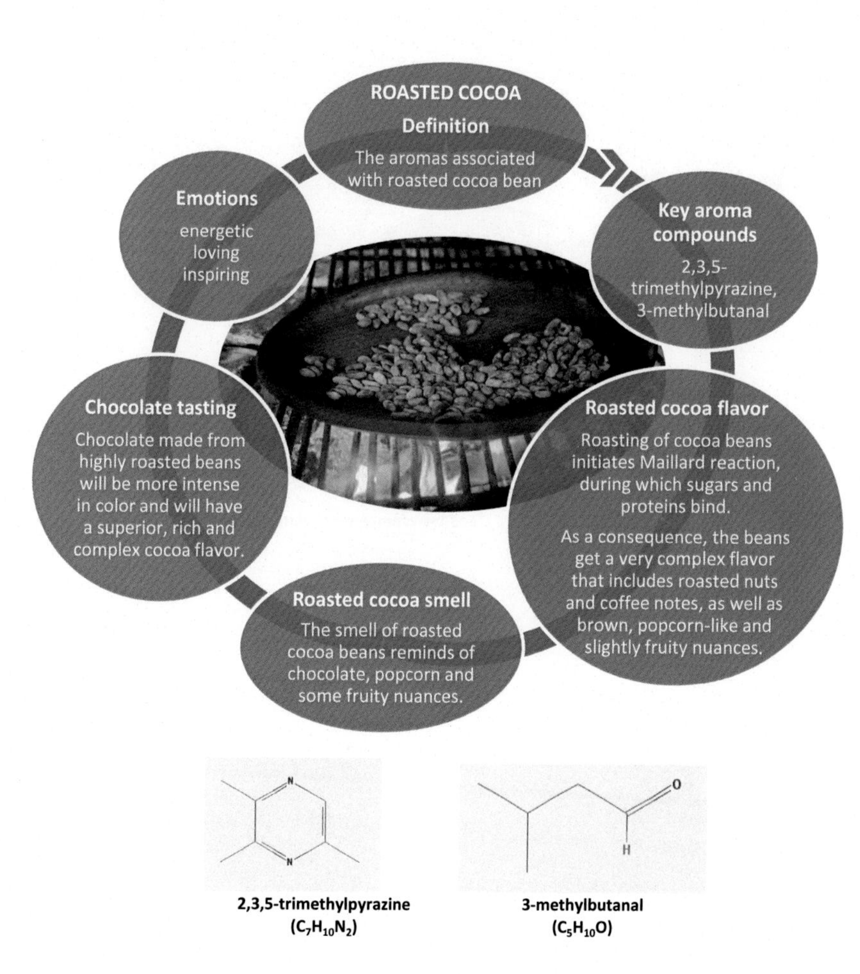

Roasted coffee

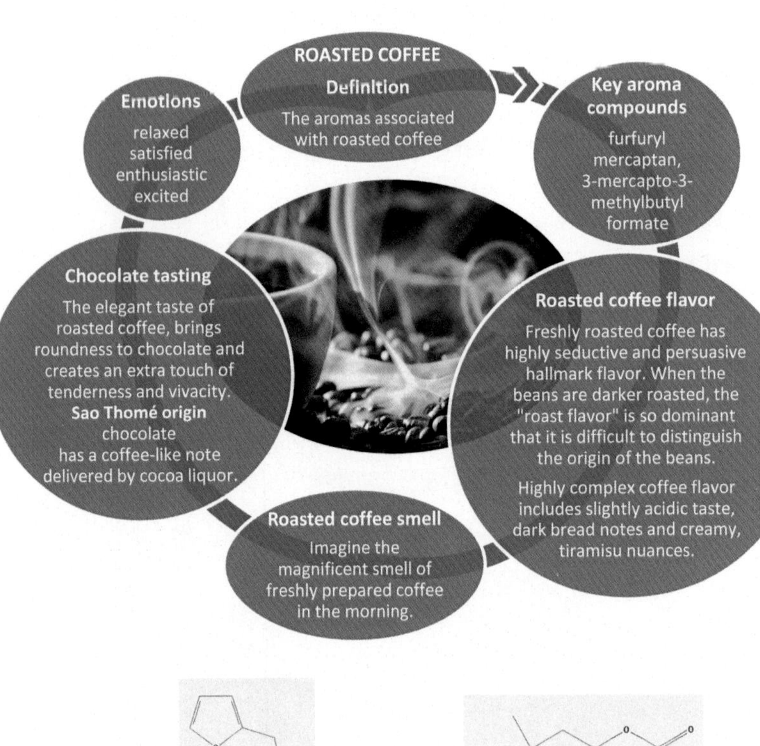

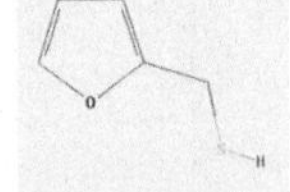

furfuryl mercaptan (C_5H_6OS)

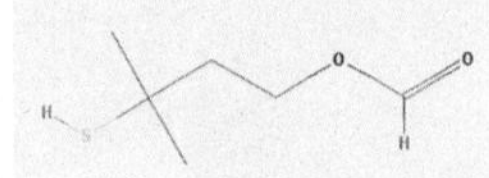

3-mercapto-3-methylbutyl formate ($C_6H_{12}O_2S$)

Toasted bread

2-acetyl-1-pyrroline (C_6H_9NO)

Roasted hazelnuts

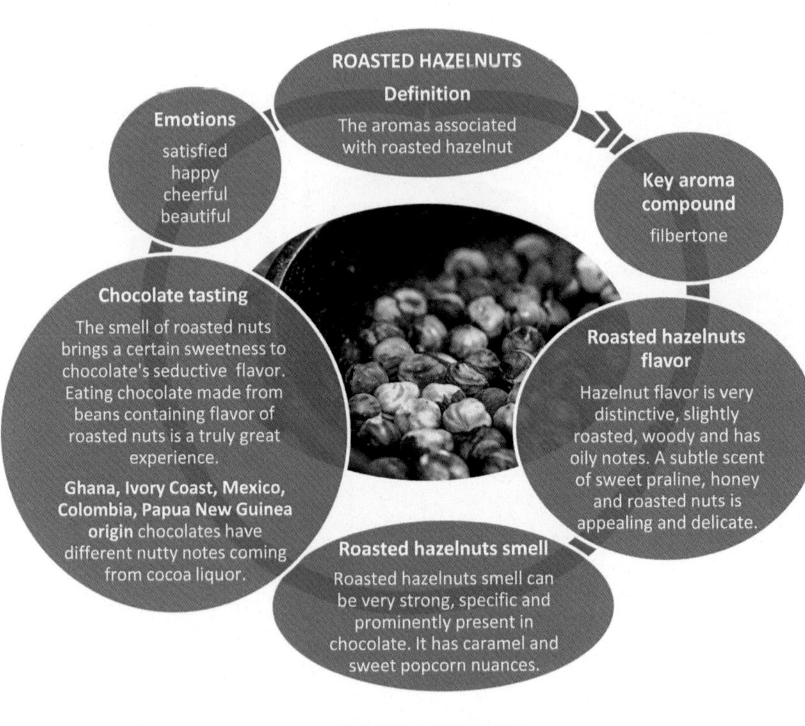

filbertone ($C_8H_{14}O$)

Roasted almonds

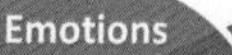

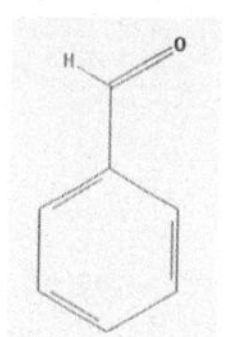

benzaldehyde (C_7H_6O)

Mokka

MOKKA

Definition

The aromas associated with whipped cream and coffee extract

Key aroma compound

furfuryl methyl disulfide

Mokka flavor

Mokka flavor is one of the most admirable scents in chocolate.

It is the flavor found in products described as rich, dairy, creamy, with caramellic, coffee and other roasted notes.

Mokka smell

The wonderful mokka smell consists of a mixture of creamy, dairy, coffee notes, grilled nuts and other typical roasted aromas.

Chocolate tasting

Anyone who tasted the mokka coffee will identify this flavor in chocolate.

The residents of Yemen, on the Red Sea, know exactly how much this smell contributes to the unique character of most of their coffee products.

Emotions

sensual
fun

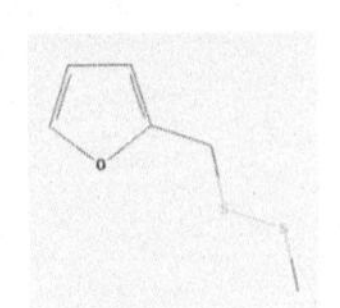

furfuryl methyl disulfide ($C_6H_8OS_2$)

GOLDEN

Honey

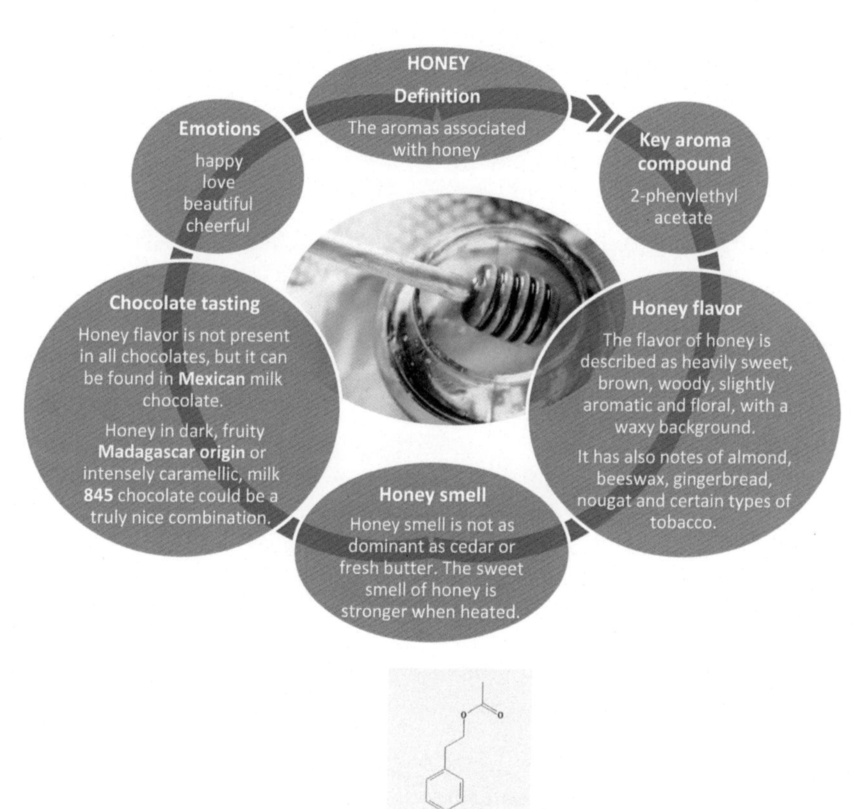

2-phenylethyl acetate ($C_{10}H_{12}O_2$)

Toffee

ethyl maltol ($C_7H_8O_3$)

furaneol ($C_6H_8O_3$)

Caramel

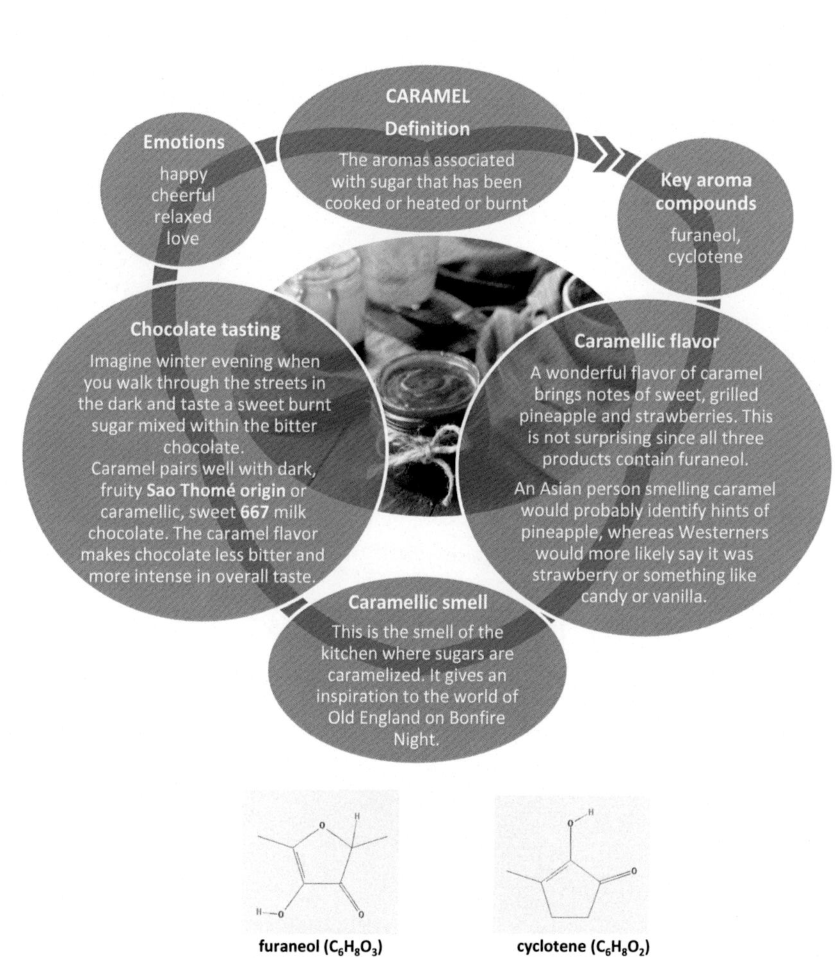

furaneol ($C_6H_8O_3$)

cyclotene ($C_6H_8O_2$)

Maple syrup

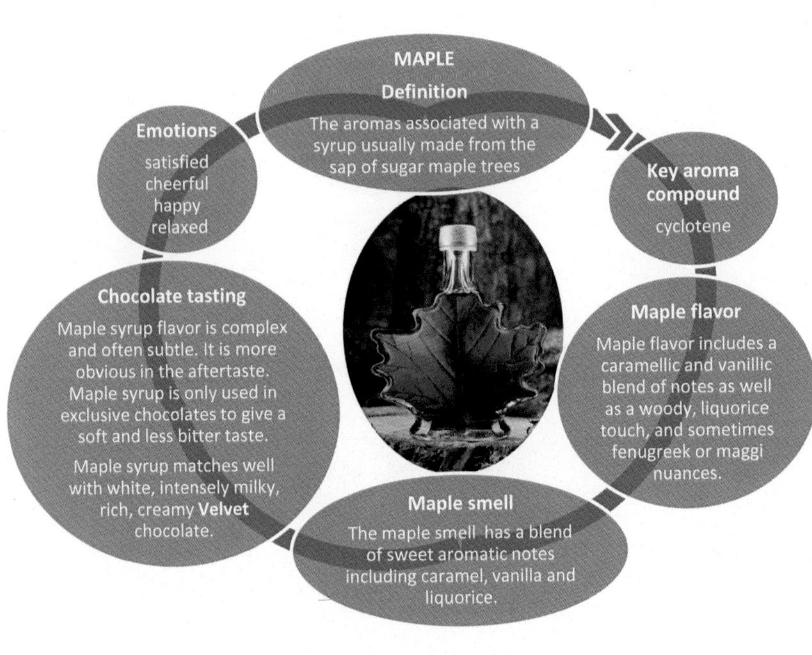

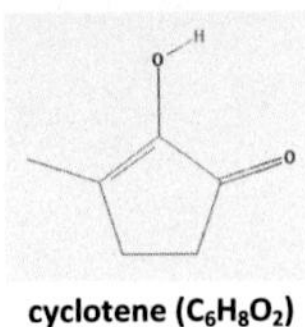

cyclotene ($C_6H_8O_2$)

GRAINS AND CEREALS

Malty

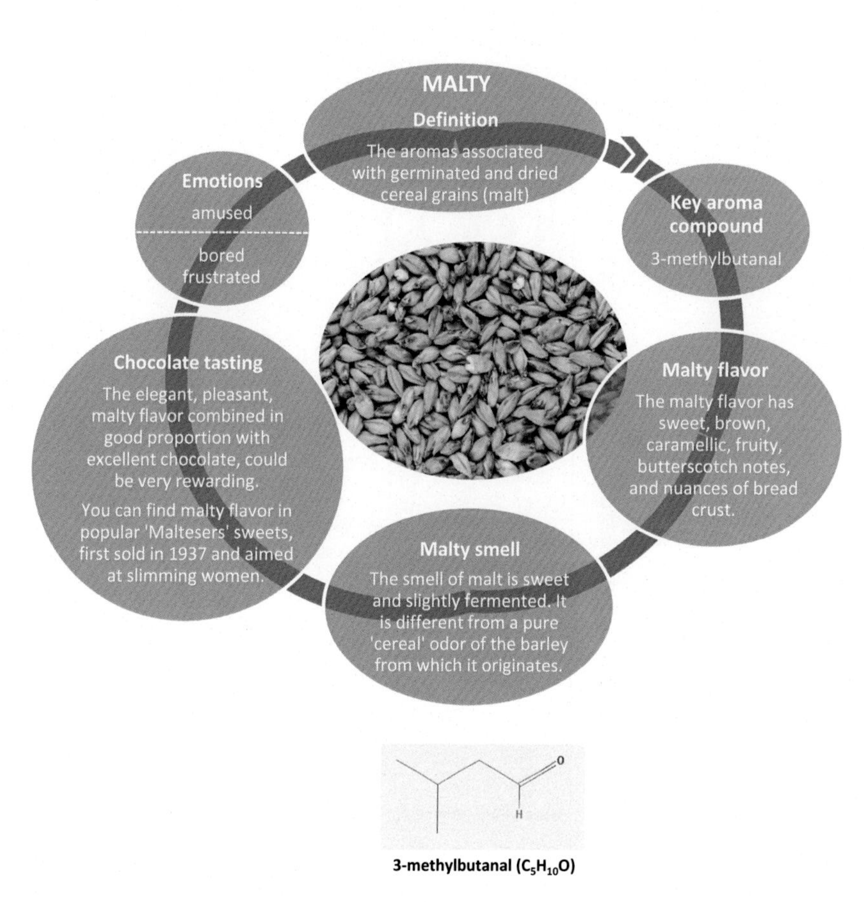

3-methylbutanal ($C_5H_{10}O$)

Sweet popcorn

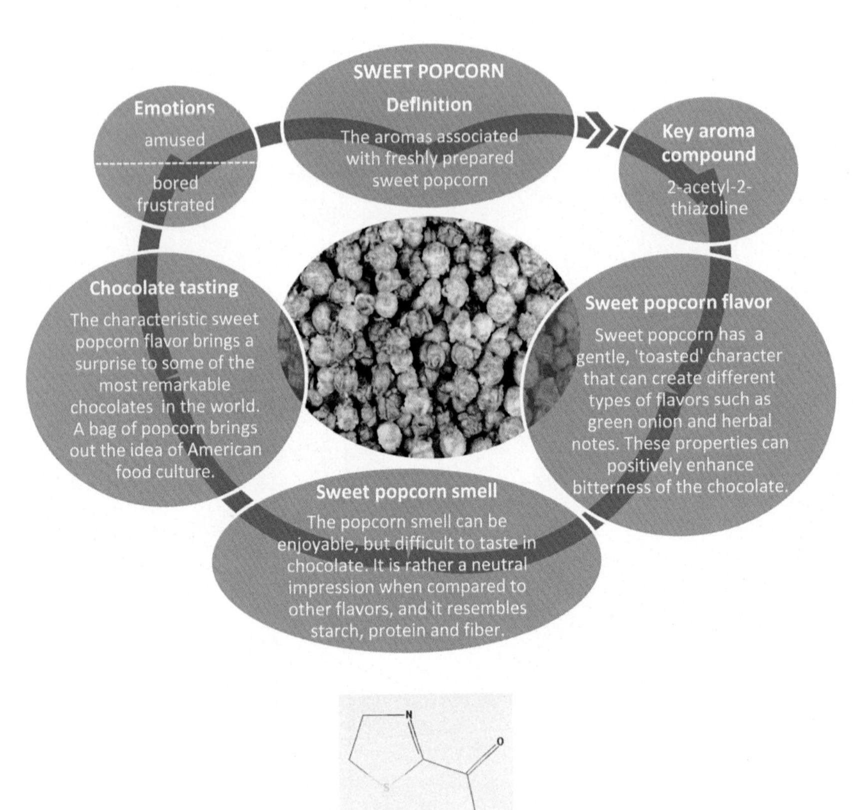

2-acetyl-2-thiazoline (C_5H_7NOS)

Cereal

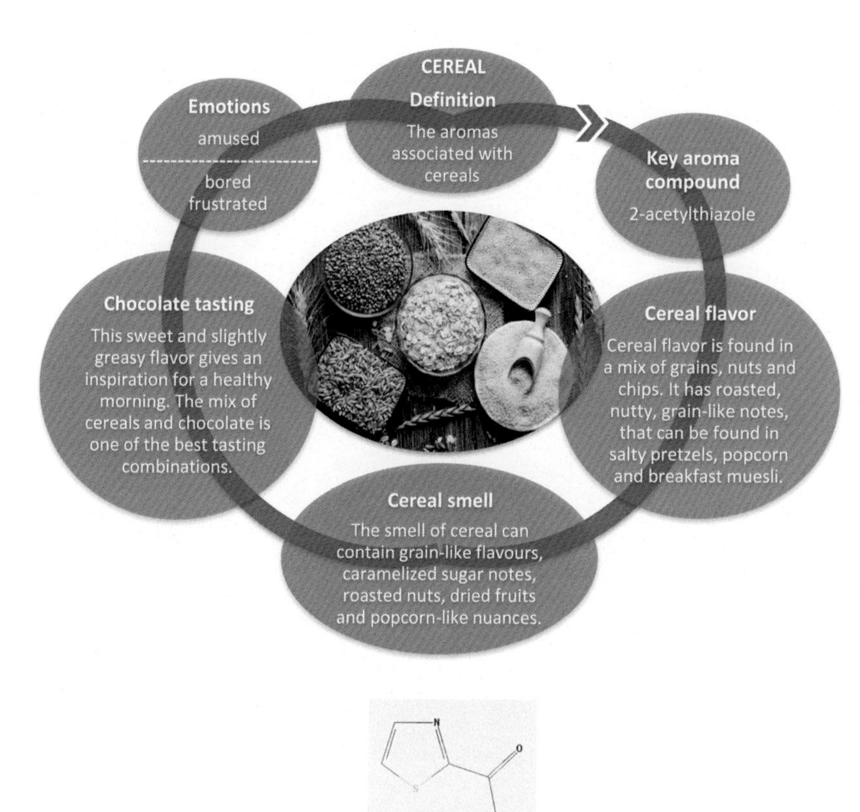

2-acetylthiazole (C_5H_5NOS)

Biscuity

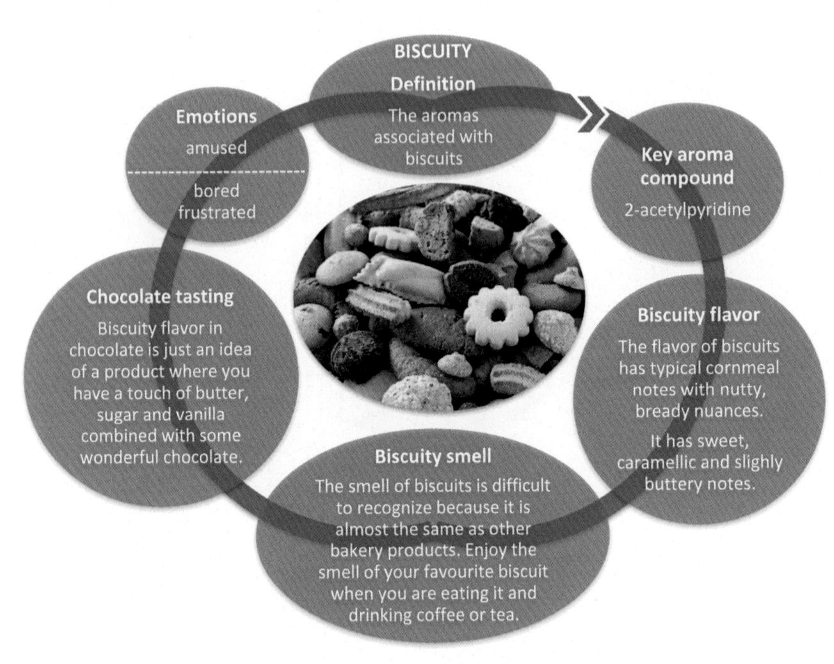

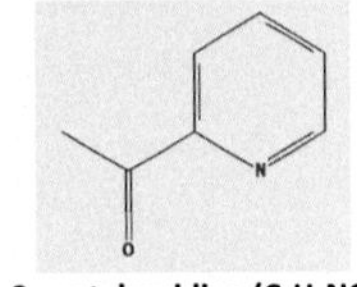

2-acetylpyridine (C_7H_7NO)

NUTTY

Hazelnut

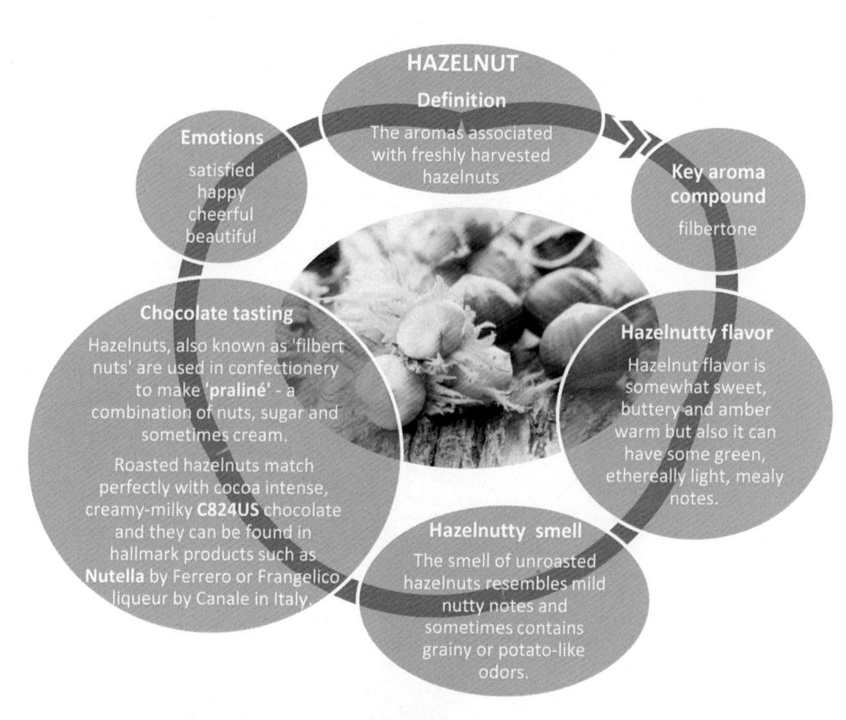

Hazelnuts surrounded by the outer, fibrous husk and a smooth shell

filbertone ($C_8H_{14}O$)

Almond

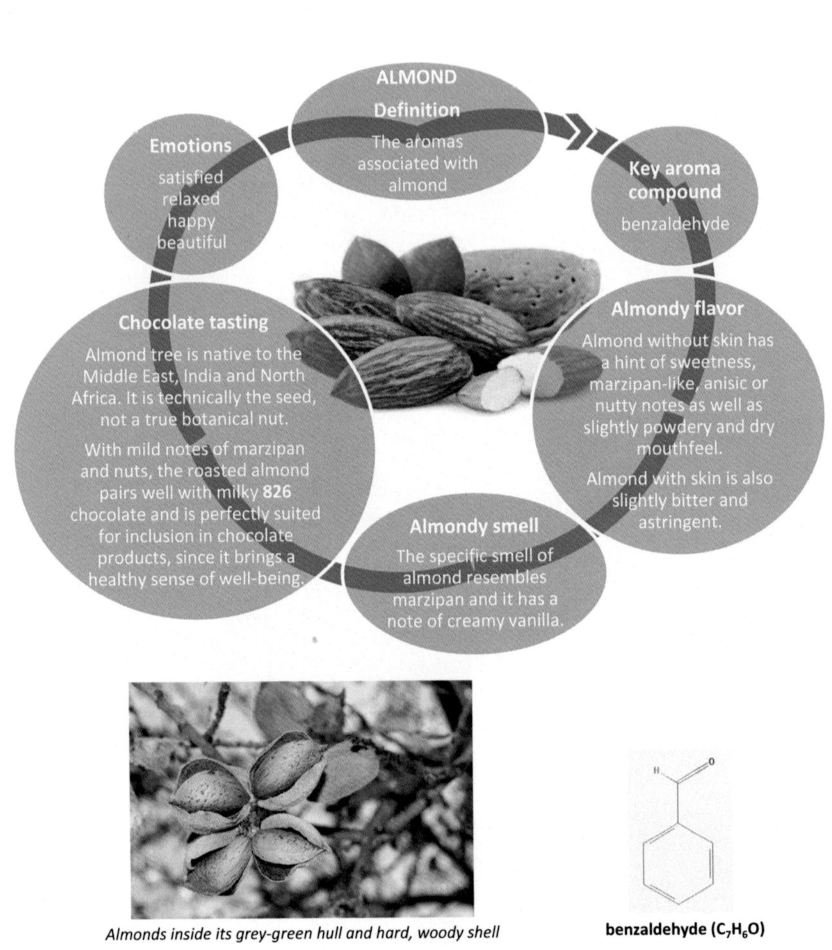

Almonds inside its grey-green hull and hard, woody shell

benzaldehyde (C_7H_6O)

Pistachio

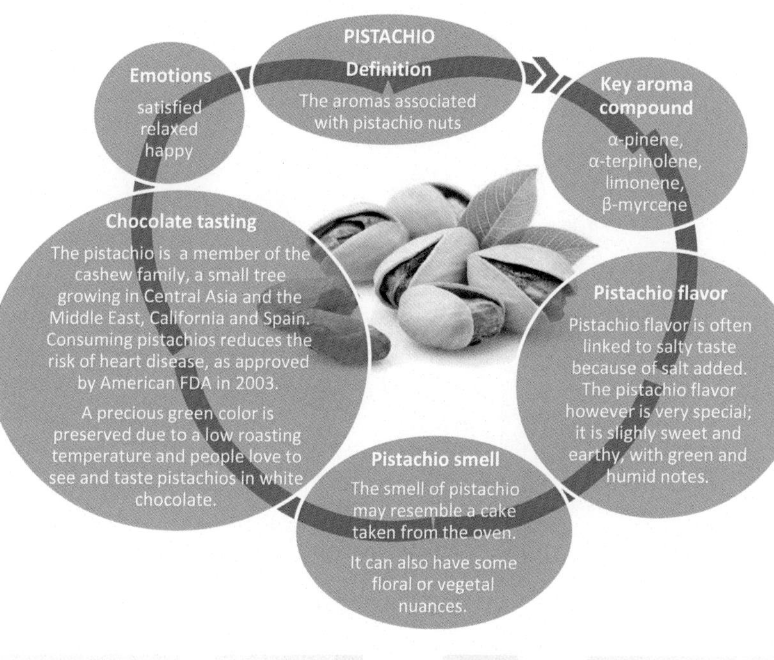

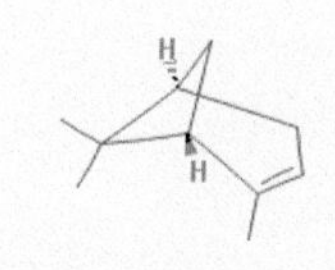

α-pinene ($C_{10}H_{16}$) (15.53–48.57%),

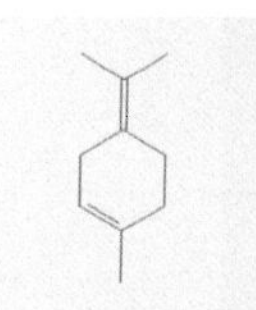

α-terpinolene ($C_{10}H_{16}$) (1.66–23.06%)

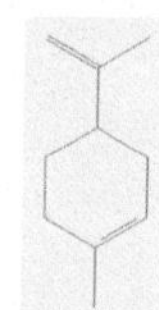

limonene ($C_{10}H_{16}$) (3.15–30.04%)

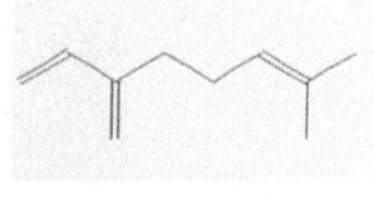

β-myrcene ($C_{10}H_{16}$) (3.50–8.95%)

Pistachio fruits ripening

Macadamia

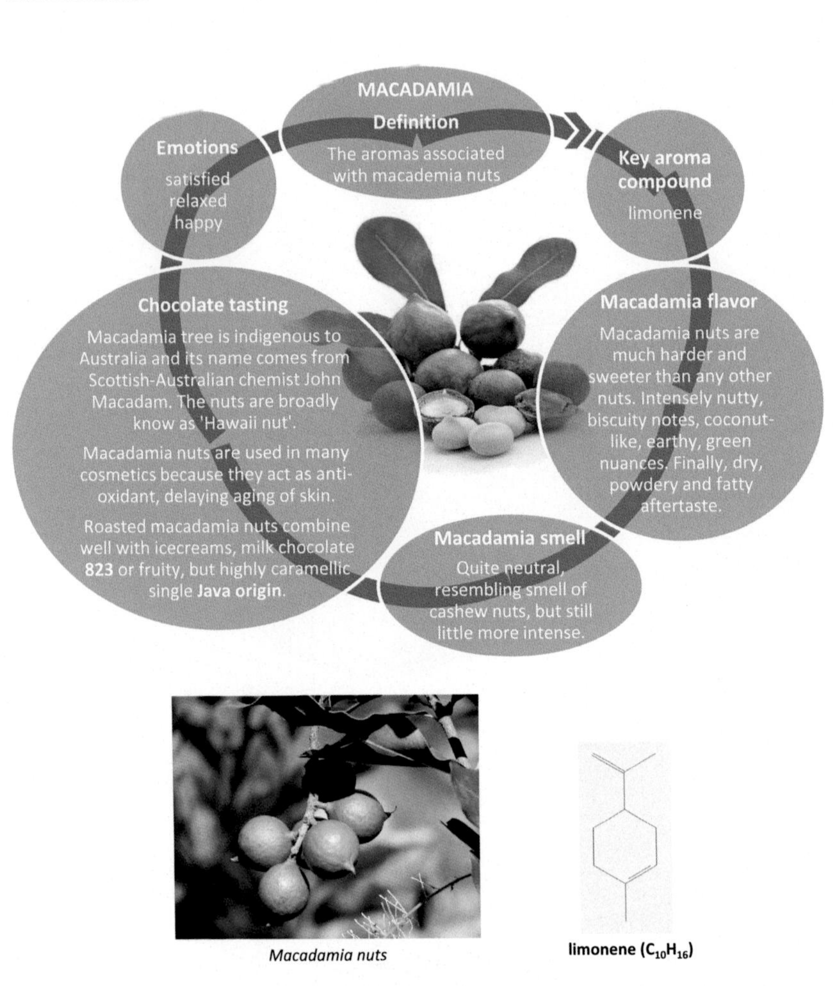

Macadamia nuts

limonene ($C_{10}H_{16}$)

Walnut

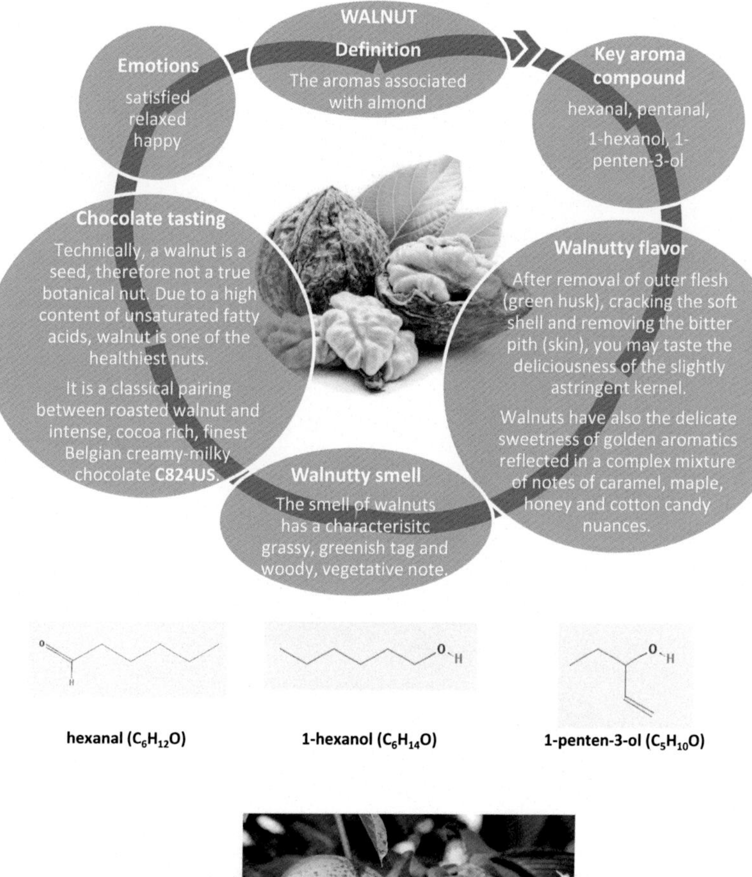

Walnut inside its green huskand hard shell

Pecan

PECAN

Definition

The aromas associated with pecan nuts

Key aroma compound

2-acetyl-1-pyrroline, 2-propionyl-1-pyrroline

Pecan flavor

Pecan has a well balanced flavor composed of sweet taste and buttery notes. It is also dry, soft and crumby in the mouthfeel, slightly astringent in the end.

Pecan smell

Slightly spicy, paprika/pepper-like smell, sweet like maple, earthy and very complex, intense.

Chocolate tasting

A large pecan tree is native to Mexico, but also growing in Texas. A major ingredient in praline candy, the pecan is not a nut but technically a fruit with a single stone or pit.

Matching roasted pecan with dark, fruity **Ecuador origin** chocolate - having notes of rum, whisky and tobacco - is a true delicacy.

Emotions

satisfied
relaxed
happy

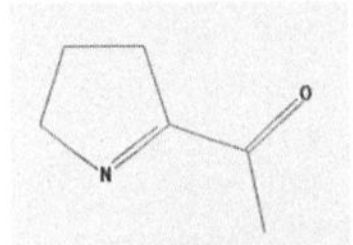

2-acetyl-1-pyrroline (C_6H_9NO)

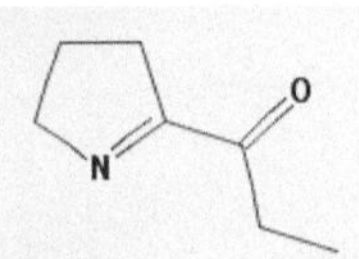

2-propionyl-1-pyrroline ($C_7H_{11}NO$)

Ripe pecan nuts

Cashew

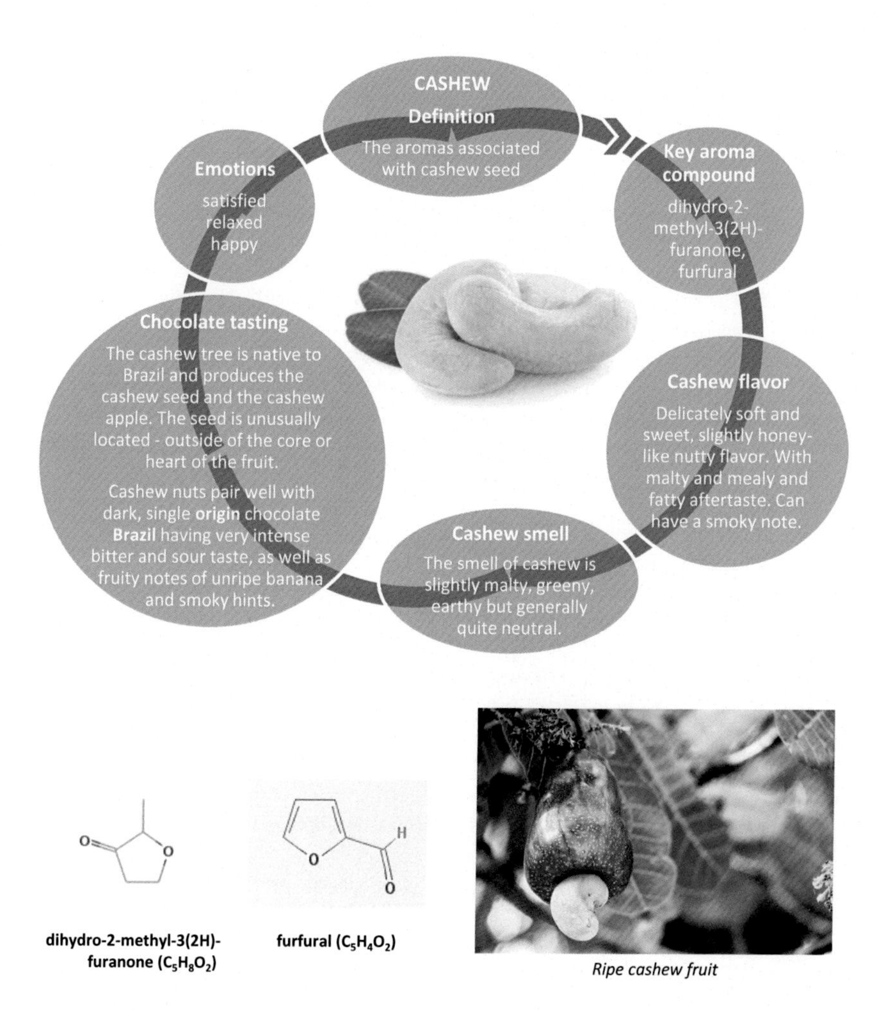

dihydro-2-methyl-3(2H)-furanone ($C_5H_8O_2$)

furfural ($C_5H_4O_2$)

Ripe cashew fruit

FURTHER READING

Afoakwa, E.O., 2010. Chocolate Science and Technology. Wiley-Blackwell, Oxford, (receptors, roasted cocoa).

Akinhanmi, T.F., Atasie, V.N., Akintokun, P.O., 2008. Chemical composition and physicochemical properties of cashew nut (Anacardium occidentale) oil and cashew nut shell liquid. J. Agric. Food Environ. Sci. 2 (1), 1–10, (cashew).

Alasalvar, C., Shahidi, F., 2008. Tree Nuts. Composition, Phytochemicals and Health Effects. Neutraceutical Science and Technology. CRC Press. Taylor & Francis Group, Boca Raton, (macadamia).

Alasalvar, C., Shahidi, F., Cadwallader, K.R., 2003. Comparison of natural and roasted Turkish tombul hazelnut (*Corylus avellana* L.) volatiles and flavor by DHA/GC/MS and descriptive sensory analysis. J. Agric. Food Chem. 51 (17), 5067–5072, (hazelnuts).

Belitz, H.D., Grosch, W., Schieberle, P., 2009. Food Chemistry. Springer Science & Business Media, Germany, (almonds, coffee, bread, buttery, cereal, sweet popcorn).

Berglund, P.T., Fastnought, C.E., Holm, E.T., 1994. Physicochemical and sensory evaluation of extruded high-fiber barley cereals. Cereal Chem. 71 (1), 91–95, (cereal).

Cadwallader, K.R., Kim, H., Puangpraphant, S., Lorjaroenphon, Y., 2010. Changes in the aroma components of pecans during roasting. In: Blank, R.I., Wüst, M., Yeretzian, C. (Eds.), Expression of Multidisciplinary Flavor Science. Zürcher Hochschule für Angewandte Wissenschaften, Winterthur, Schweizpp. 301–304, (pecan).

Caporaso, N., Genovese, A., Canela, M.D., Civitella, A., Sacchi, R., 2014. Neapolitan coffee brew chemical analysis in comparison to espresso, mokka and American brews. Food Res. Int. Vol. 61, 152–160, (mokka).

Chang, C.-Y., Seitz, L.M., Chambers, I.V.E., 1995. Volatile flavor components of breads made from hard red winter wheat and hard white winter wheat. Cereal Chem. 72 (3), 237–242, (bread).

Civille, G.V., Lapsley, K., Huang, G., Yada, S., Seltsam, J., 2010. Development of an almond lexicon to assess the sensory properties of almond varieties. J. Sens. Stud. 25 (1), 146–162, (almonds).

Cˇížková, H., Rajchl, A., Šnebergrová, J., Voldřich, M., 2013. Filbertone as a marker for the assessment of hazelnut spread quality. J. Food Sci. 31 (1), 81–87, (hazelnuts).

Coghe, S., Martens, E., D'Hollander, H., Dirinck, P.J., Delvaux, F.R., 2012. Sensory and instrumental flavor analysis of wort brewed with dark specialty malts. J. Inst. Brew. 110 (2), 94–103, (malty).

Czerny, M., Mayer, F., Grosch, W., 1999. Sensory study of the character impact odorants of roasted Arabica coffee. J. Agric. Food Chem. 47 (2), 695–699, (coffee).

He, S., Hekmat, S., 2015. Sensory evaluation of non-dairy probiotic beverages. J. Food Res. 4 (1), 186–192, (almond).

Jing, W., Liu, Y.-I., Peng, S.-D., Lin, L.-J., Li, J.-H., 2016. Effects of roasting on volatile components profile in macadamia nut kernels (Macadamia Integrifolia). Guangdong Agric. Sci. 43 (7), 110–119, (macadamia).

Kendirci, P., Altuğ Onoğur, T., 2011. Investigation of volatile compounds and characterization of flavor profiles of fresh pistachio nuts (*Pistacia vera* L). Int. J. Food Prop. 14 (2), 319–330, (pistachio).

Kumazawa, K., Masuda, H., 2003. Identification of odor-active 3-mercapto-3-methylbutyl acetate in volatile fraction of roasted. Coffee brew isolated by steam distillation under reduced pressure. J. Agric. Food Chem. 51 (10), 3079–3082, (coffee).

Liu, Z.-H., Liu, C.-B., Chen, Y.-K., He, X.-Y., Sun, Z.-Y., Miao, M.-M., 2013. Determination of volatile organic compounds in Macadamia flower by simultaneous distillation extraction and static headspace. Asian J. Chem. 25 (3), 1265–1269, (macadamia).

Marsili, R., 2001. Flavor, Fragrance and Odor Analysis. Marcel Dekker, Inc, New York, (mokka).

Miller, A.E., Chambers, I.V.E., Jenkins, A., Lee, J., Chambers, D.H., 2003. Defining and characterizing the "nutty" attribute across food categories. Food Qual. Pref. 27 (1), 1–7, (hazelnuts).

Nazaruddin, R., Osman, H., Mamot, S., Wahid, S., Nor Aini, I., 2006. Influence of roasting conditions on volatile flavor of roasted Malaysian cocoa beans. J. Food Process. Preserv. 30 (3), 280–298, (roasted cocoa).

Nursten, H.E., 1997. The flavor of milk and dairy products: I. Milk of different kinds, milk powder, butter and cream. Int. J. Dairy Technol. 50 (2), 48–56, (toffee).

ONP, 2003. Office of Nutritional Products, Labeling and Dietary Supplements (23 July 2003)."Qualified Health Claims: Letter of Enforcement Discretion – Nuts and Coronary Heart Disease (Docket No 02P-0505)". Center for Food Safety and Applied Nutrition. Archived from the original on 17 June 2008. Retrieved 17 June 2008. (pistachio, FDA health claim).

Paravisini, L., Prot, A., Gouttefangeas, C., Moretton, C., Nigay, H., Dacremont, C., Guichard, E., 2015. Characterisation of the volatile fraction of aromatic caramel using heart-cutting multidimensional gas chromatography. Food Chem. 167, 281–289, (caramel).

Perkins, T.D., Van den Berg, A.K., 2009. Chapter 4 Maple Syrup—production, composition, chemistry, and sensory characteristics. Adv. Food Nutr. Res. 56, 101–143, (maple).

Piggott, J., 2012. Alcoholic Beverages: Sensory Evaluation and Consumer Research. Woodhead Pub, Oxford, (honey).

Saez, A., Montoya, S., Cabrera, J., Asensio, C., Ortega, E., 2014. Characterisation and lipid profile of Macadamia nuts (macadamia Integrifolia and Macadamia Tetraphyllia). Int. J. Eng. Appl. Sci. 4 (9), 33–39, (macadamia).

Stark, T., Bareuther, S., Hofmann, T., 2005. Sensory-guided decomposition of roasted cocoa nibs (*Theobroma cacao*) and structure determination of taste-active polyphenols. J. Agric. Food Chem. 53 (13), 5407–5418, (roasted cocoa).

Sunarharuma, W.B., Williams, D.J., Smyth, H.E., 2014. Complexity of coffee flavor: a compositional and sensory perspective. Food Res. Int. 62, 315–325, (coffee, malty).

Tarancón, P., Fiszman, S.M., Salvador, A., Tárrega, A., 2013. Formulating biscuits with healthier fats. Consumer profiling of textural and flavor sensations during consumption. Food Res. Int. 53, 134–140, (biscuity).

Tran, P.D., Van de Walle, D., De Clercq, N., De Winne, A., Kadow, D., Lieberei, R., Messens, K., Tran, D.N., Dewettinck, K., Van Durme, J., 2015. Assessing cocoa aroma quality by multiple analytical approaches. Food Res. Int. 77, 657–669, (sweet popcorn).

leff www.leffingwell.com. (caramel, maple).

Xiao, L., Lee, J., Zhang, G., Ebeler, S.E., Wickramasinghe, N., Seiber, J., Mitchell, A.E., 2014. HS-SPME GC/MS characterization of volatiles in raw and dry-roasted almonds (Prunus dulcis). Food Chem. 151, 31–39, (almonds).

Zhou, Q., Wintersteen, K.L., Cadwallader, K.R., 2002. Identification and quantification of aroma-active components that contribute to the distinct malty flavor of buckwheat honey. J. Agric. Food Chem. 50 (7), 2016–2021, (honey).

Chapter 4

Dairy

Chapter Outline

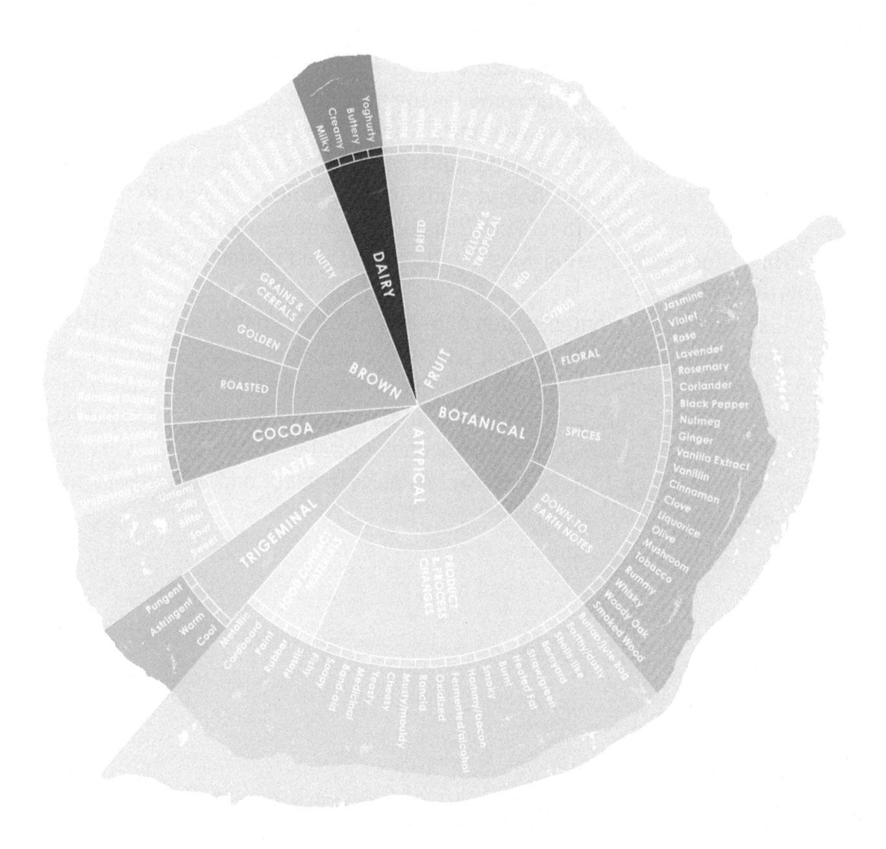

Hidden Persuaders in Cocoa and Chocolate. http://dx.doi.org/10.1016/B978-0-12-815447-2.00005-X

In the production of white and milk chocolate, the cocoa products (liquor, powder and butter) are mixed with sugar and milk powder.

Evaporating milk creates milk powder that is quite rich in milk fat, proteins and carbohydrates. The amount of fat in milk powders influences initial creaminess. However, not only fat is important for sensory perception: proteins and carbohydrates also play a role.

Milk contains several milk proteins which are biologically active, including lactoferrin, lactoperoxidase and cyastin. Apart from fat and protein, milk is rich in carbohydrates, such as lactose, glucose, galactose and other oligosaccharides. The lactose gives milk its 'sweet' taste, which is further influenced by the type of milk powder preparation.

In chocolate, three kinds of milk powder are used: whole milk, skimmed milk or whey powder. Whole milk powder is made from whole milk and gives more 'creamy' flavor while skimmed powder contains less milk fat contributing to 'milky' notes. Whey milk powder is a cheaper alternative for skimmed powder and it has a characteristic 'salty' taste that can be quite nice in special chocolate recipes.

Two major techniques are used to produce milk powder from milk: spray drying and roller drying. In the spray drying method, a dry milk powder is made from liquid milk by rapidly drying milk – sprayed as droplets – with a hot gas, which preserves thermally sensitive milk components and accentuates 'milky' flavors. In the roller dried technique, milk is continuously brought to the heated surface of the rollers where it is dried, leading to a more complex 'caramel' flavor. The roller method, using whole milk, is practiced in small dairy companies and is disappearing.

Milk fat and cocoa butter match extremely well in contributing to softening the bitter compounds of the cocoa part and creating a unique sensory experience. By playing with fat, protein and sugar composition we can create various sensory profiles of milk chocolate. It shall be remembered that only nice dairy derived flavors are appreciated in chocolate and they include: *milky*, *creamy*, *yoghurty* and *buttery*. Finally, 'cheesy' means off-flavor and is discussed in the last chapter.

MILKY

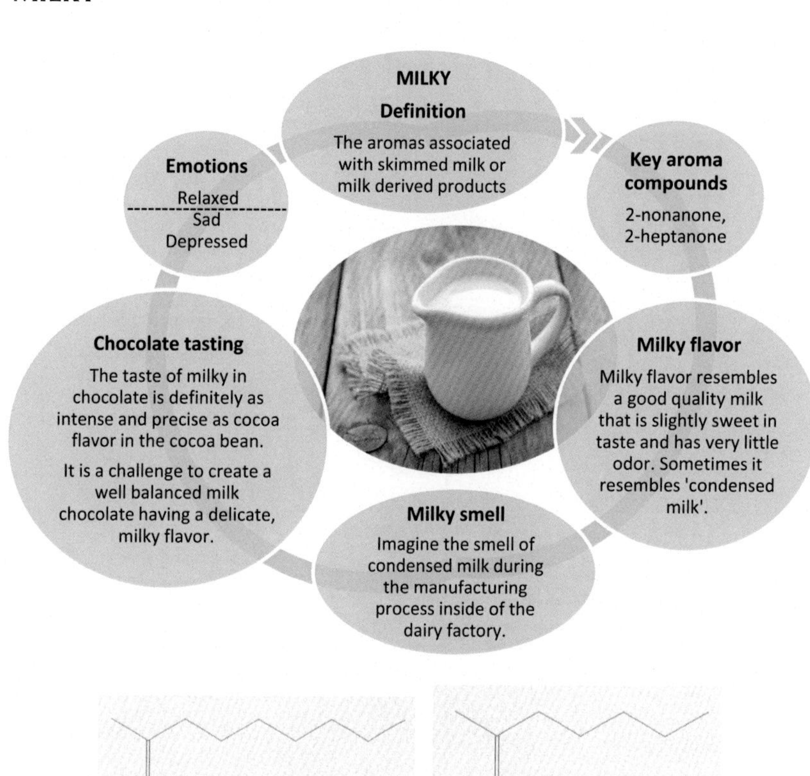

2-nonanone ($C_9H_{18}O$)

2-heptanone ($C_7H_{14}O$)

CREAMY

CREAMY

Definition

A sweet, dairy note associated with cream or other highly fatty dairy products

Key aroma compound

Acetoin

Creamy flavor

Creaminess is understood as mouthfeel (texture) and milky flavor, accompanied with strong buttery nuances, giving a complex, smooth and rich feeling.

It is often associated with a general lactonic note found in dairy products, peach and apricot.

Creamy smell

The smell of creaminess displays rich, dairy aromatic compounds and brings 'body' to the chocolate.

Chocolate tasting

The creamy flavor in chocolate is a synonym for 'rich taste', thick mouthfeel or complex texture, mildness and some sensual intensity.

Emotions

Happy
Satisfied
Relaxed
Love

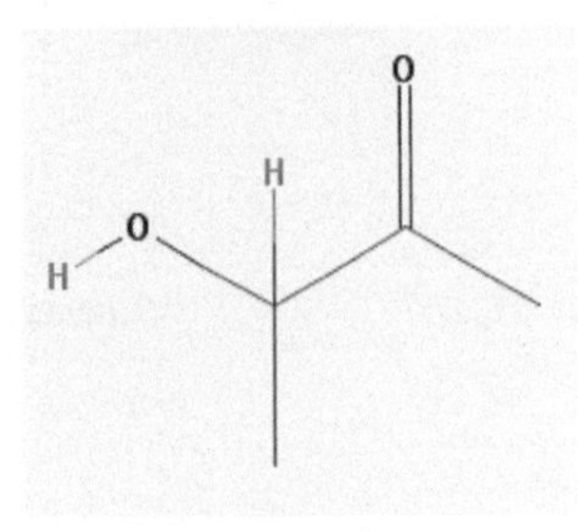

acetoin ($C_4H_8O_2$)

BUTTERY

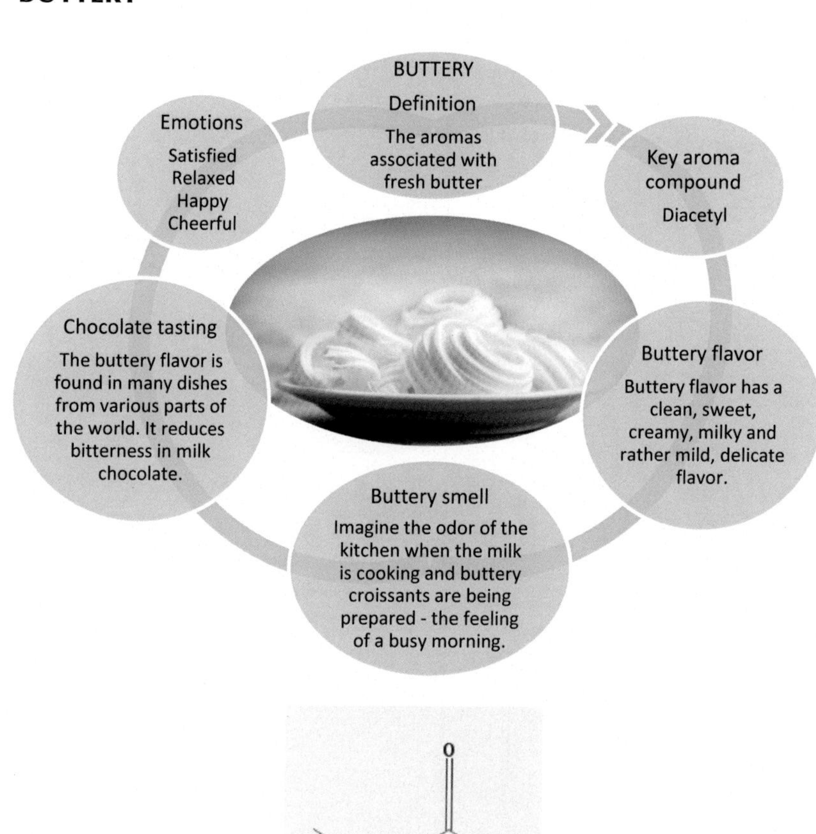

diacetyl ($C_4H_6O_2$)

YOGHURTY

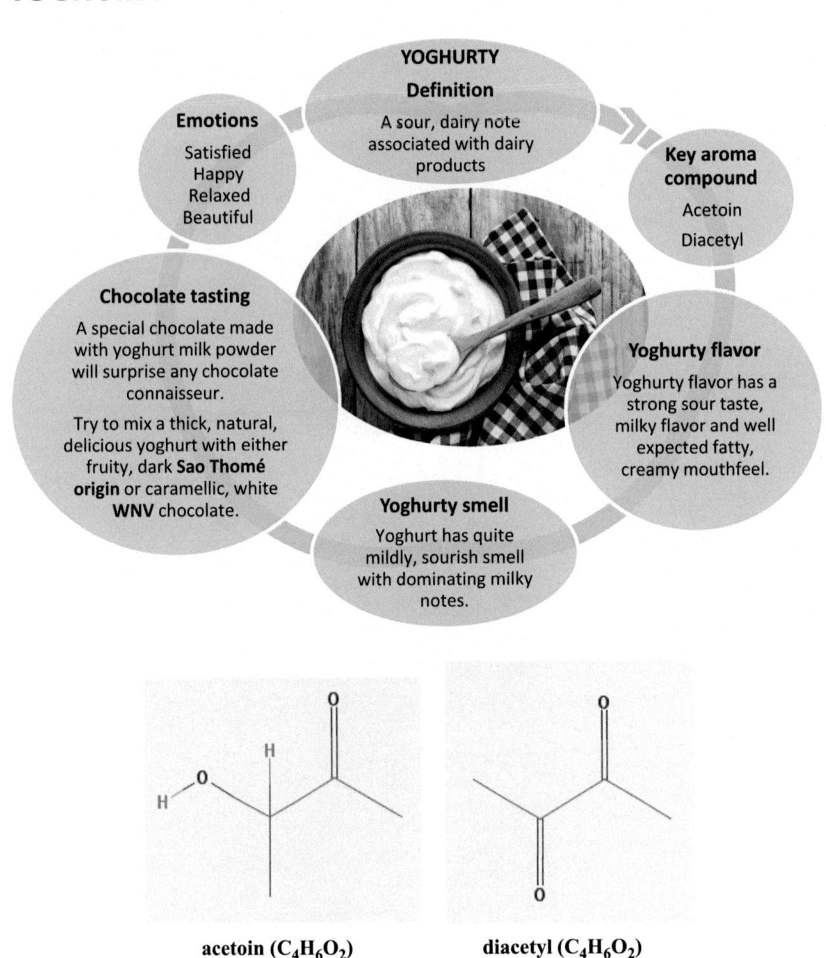

acetoin ($C_4H_6O_2$) **diacetyl ($C_4H_6O_2$)**

FURTHER READING

Chartier, F., 2012. Taste Buds and Molecules: The Art and Science of Food, Wine and Flavor. John Wiley & Sons, Inc, New Jersey, (creamy).

Cheng, H., 2010. Volatile flavor compounds in yoghurt: a review. Crit. Rev. Food Sci. Nutr. 50 (10), 938–950, (yoghurty).

Friedrich, J.E., Acree, T.E., 1998. Gas chromatography olfactometry (GC/O) of dairy products. Int. Dairy J. 8 (3), 235–241, (milky).

Liang, B., Hartel, R.W., 2004. Effects of milk powders in milk chocolate. J. Dairy Sci. 87 (1), 20–31, (processing techniques).

Singh, T.K., Drake, M.A., Cadwallader, K.R., 2003. Flavor of cheddar cheese: a chemical and sensory perspective. Compr. Rev. Food Sci. Food Saf. 2, 166–189, *buttery*.

Stampanoni, C.R., 1994. The use of standardized flavor languages and quantitative flavor profiling technique for flavored dairy products. J. Sens. Stud. 9 (4), 383–400, *buttery*.

Chapter 5

Fruit

Chapter Outline

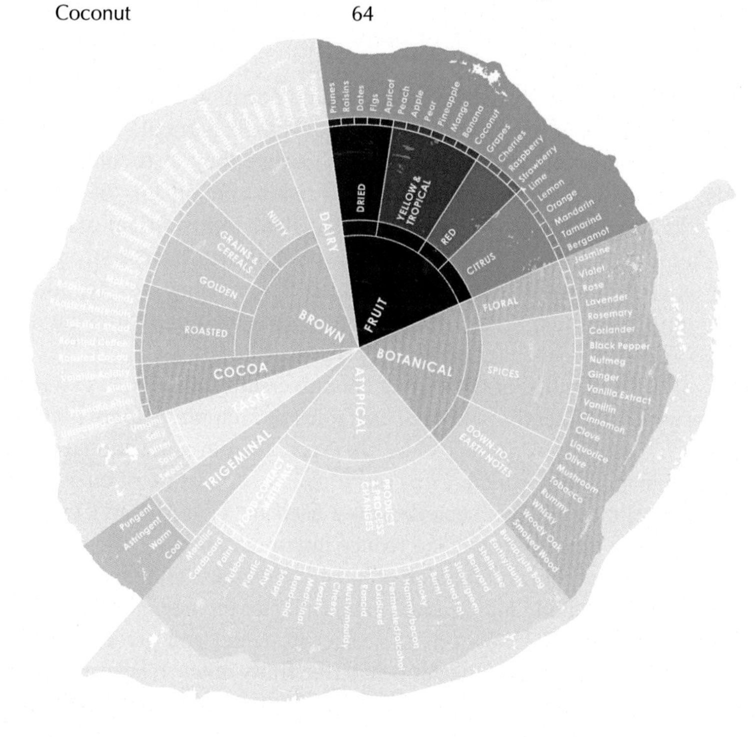

Hidden Persuaders in Cocoa and Chocolate. http://dx.doi.org/10.1016/B978-0-12-815447-2.00006-1

For many people, fruity notes represent all kinds of warm 'Spring and Summer' sensory impressions. Fruity notes are linked to perception of top or medium volatile chemical compounds. Stability of these compounds is linked to each step of the cocoa processing and finally chocolate production. As fruitiness can be easily destroyed, full attention is given in the business to share knowledge and skills on how to preserve it.

Natural fruity notes come from good quality cocoa beans. A precious fruitiness is linked to the origin of the beans. Our extended knowledge about the sensory profiles of cocoa beans growing all over the world shows, for instance, that beans growing in Madagascar or Sao Thomé have red and yellow fruity notes, while Ecuador has a fruity note combined with nuances of rum, whisky and tobacco.

A specific fruity 'bouquet' in chocolate is perceived more easily in dark than milk or white chocolate. As the fruitiness is so delicate, it can be easily overwhelmed by other ingredients, such as powerful dairy, cereal or nutty aromatics, often present in milk chocolate.

A more complex combination of fruity notes can be detected only by highly trained sensory experts who shall be specially prepared for detection, classification and quantification of these notes and nuances. More specifically, the experts focus on the following four categories of fruity impressions: *dried fruits, yellow fruits, red fruits* and *citrusy* notes.

First, impression of so called 'dried fruits' refers to something what is very sweet and resembles concentrated sugar. In this book we focus on *apricot, raisins, prunes, figs, dates.*

'Dried fruits' are easily accessible, sweet and nutritious. They are an excellent alternative for sweets and everybody knows their particular 'taste'. Their sensory profiles are very rich and complex because they also contain lots of minerals. Due to the drying process, dried raisins contain lots of potassium whereas figs bring calcium. Prunes are an excellent source of antioxidants. In case of dates, they are not only naturally sweet but they have a stringy inner texture that gives a kind of caramellic aftertaste to the dishes they are added to. Dates do not make 'jumps' in the sugar level in the blood and it is the reason why dates are recommended for diabetic people.

Second, 'yellow fruits' are those that are universally perceived as sweet, but not necessarily sour. Hereby, we have examples of *banana, mango, pineapple, apple, peach, pear.* These fruits can be also special from a nutritional point of view, like pineapple that contains lots of bromelain, a mixture of enzymes that digest protein. Pineapple has been therefore used for centuries in Central and South America to treat ingestion and reduce inflammation.

Third, 'red fruits' bring impression of intense sweetness combined with sourness. The examples taking into account in this chapter include: *grapes, cherries, raspberry, strawberry.* These fruits are easier to detect and perceive, because there is a good mix of two basic tastes behind the fruity notes. It is also somewhat easy for people to associate that type of fruitiness with all kind of red berries.

Fourthly, 'citrusy' notes are linked to the majority of known citruses, including: *lemon, lime, mandarin, orange, bergamot, tamarind.* Citrusy impression is special and cannot be confused with other fruitiness. The challenge is however to distinguish an underlying typicality of citrusy notes: for instance lime or lemon will have more sour, sharp fruitiness while mandarin or orange bring a more sweet, delicate fruitiness. Luckily, bergamot, used as major ingredient in perfumes has a distinct sensory profile, and the same goes for tamarind. The fourth type of chocolate – ruby – invented by Barry Callebaut gives an impression of both red fruits and citrusy notes.

DRIED

Prunes

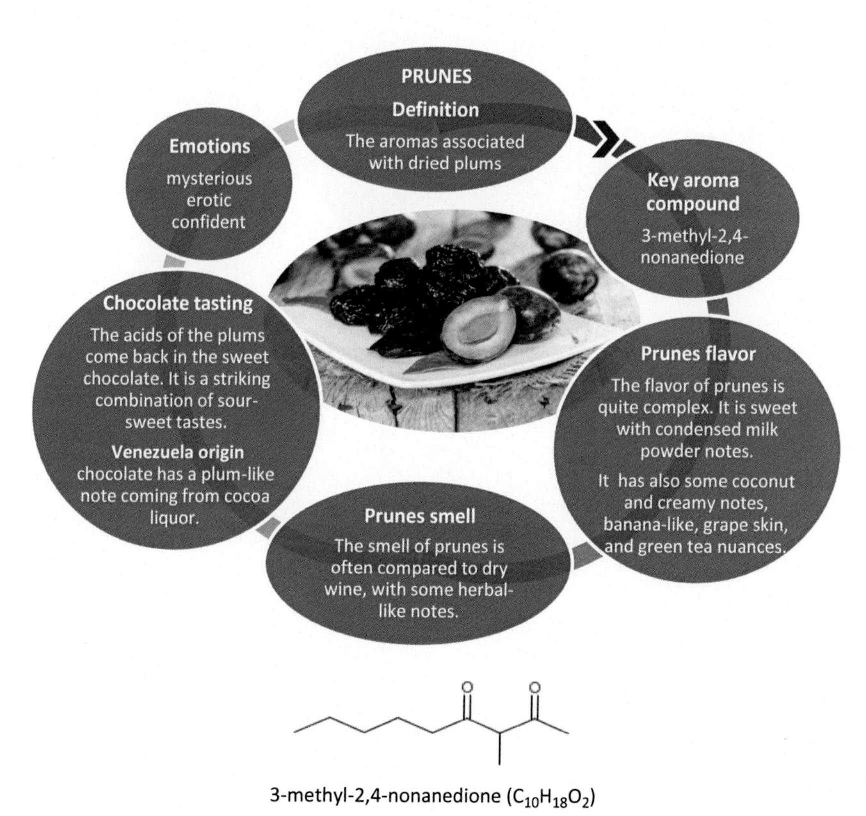

3-methyl-2,4-nonanedione ($C_{10}H_{18}O_2$)

Raisins

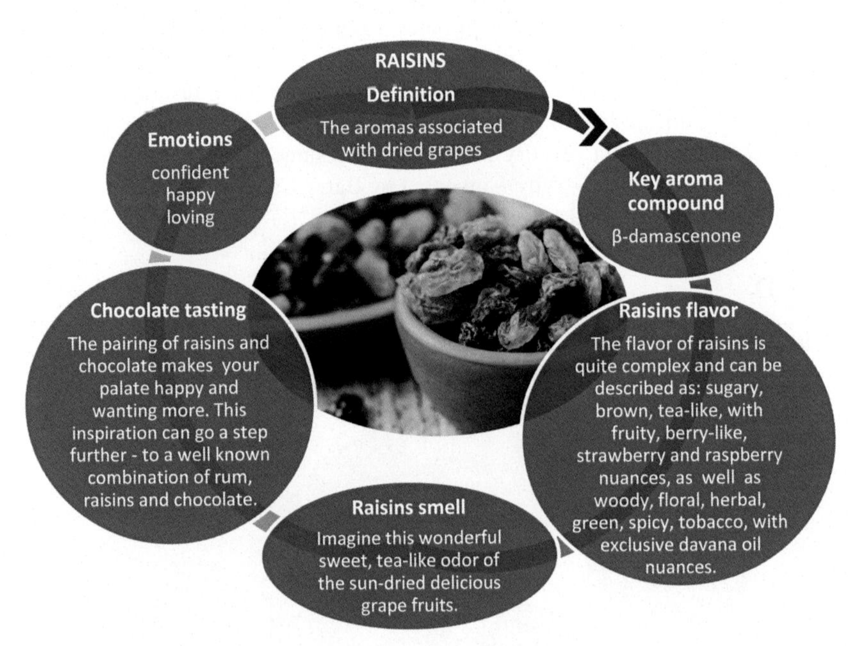

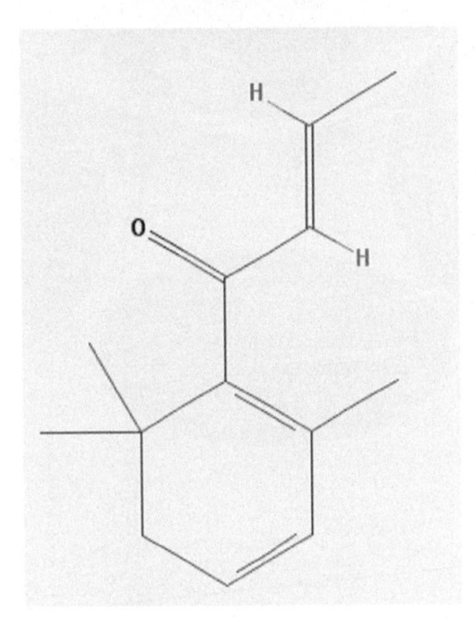

β-damascenone ($C_{13}H_{18}O$)

Dates

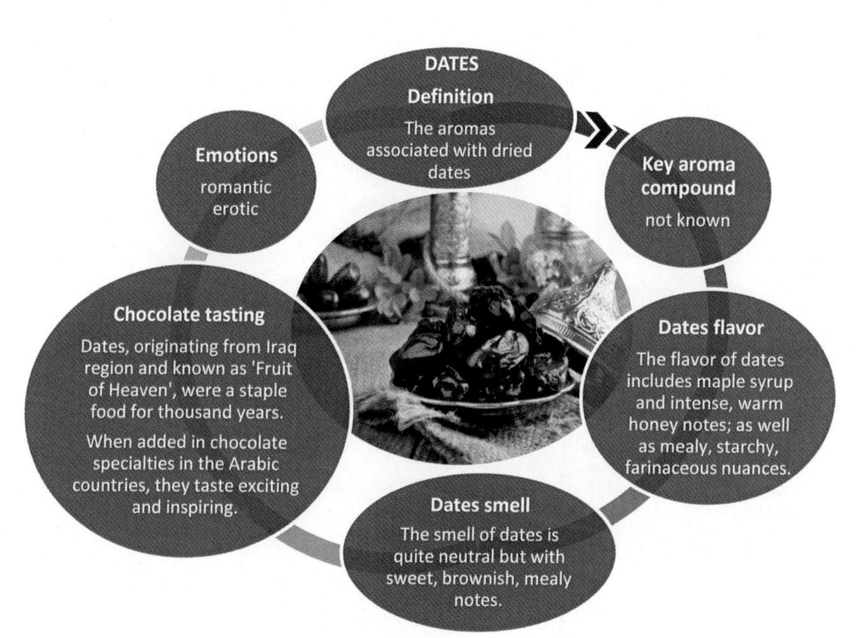
DATES
Definition
The aromas associated with dried dates
Key aroma compound
not known
Dates flavor
The flavor of dates includes maple syrup and intense, warm honey notes; as well as mealy, starchy, farinaceous nuances.
Dates smell
The smell of dates is quite neutral but with sweet, brownish, mealy notes.
Chocolate tasting
Dates, originating from Iraq region and known as 'Fruit of Heaven', were a staple food for thousand years.
When added in chocolate specialties in the Arabic countries, they taste exciting and inspiring.
Emotions
romantic
erotic

Figs

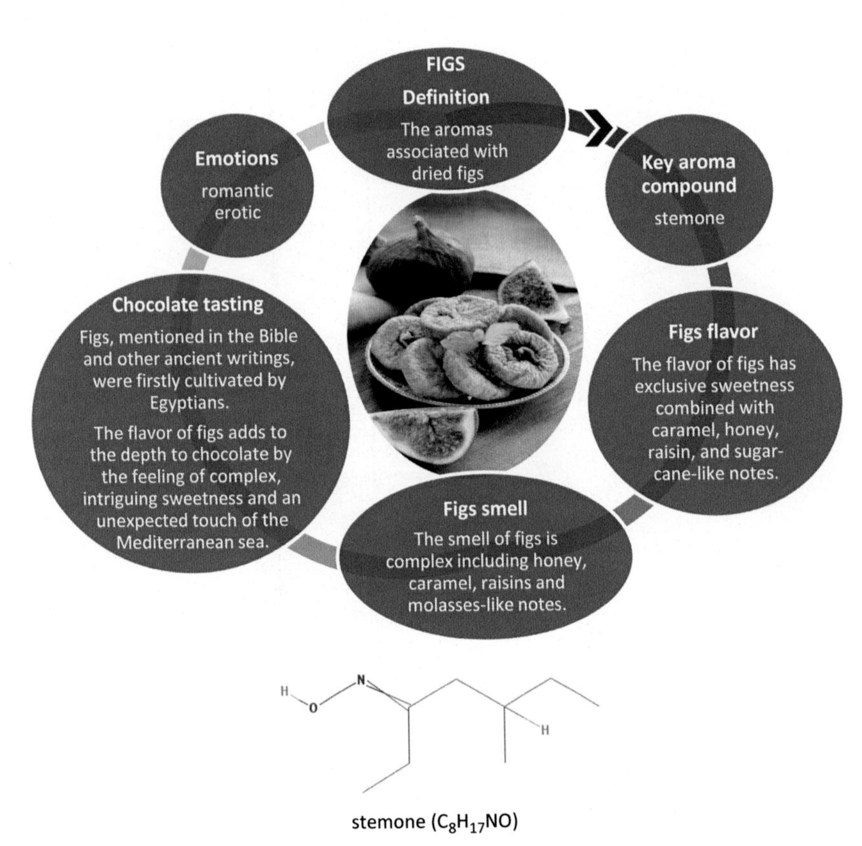

stemone ($C_8H_{17}NO$)

Apricot

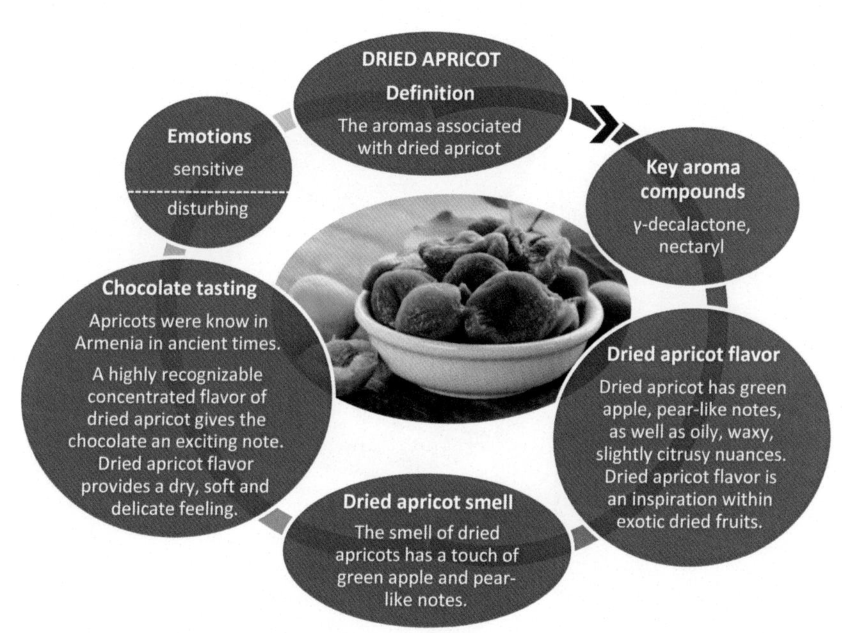

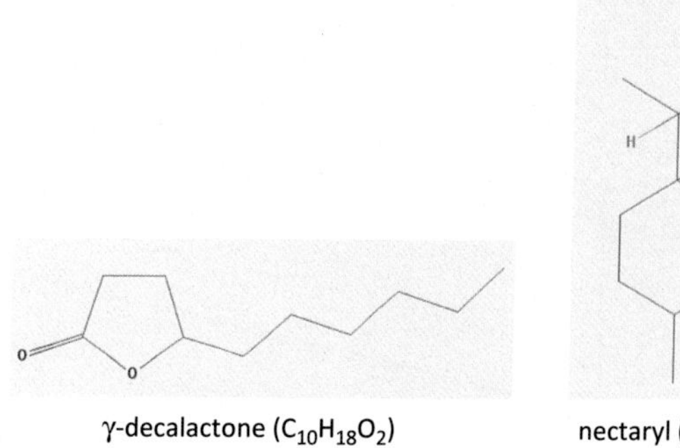

γ-decalactone ($C_{10}H_{18}O_2$)

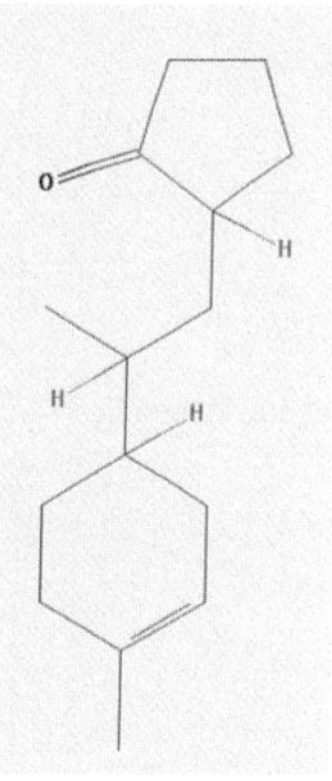

nectaryl ($C_{15}H_{24}O$)

YELLOW AND TROPICAL

Peach

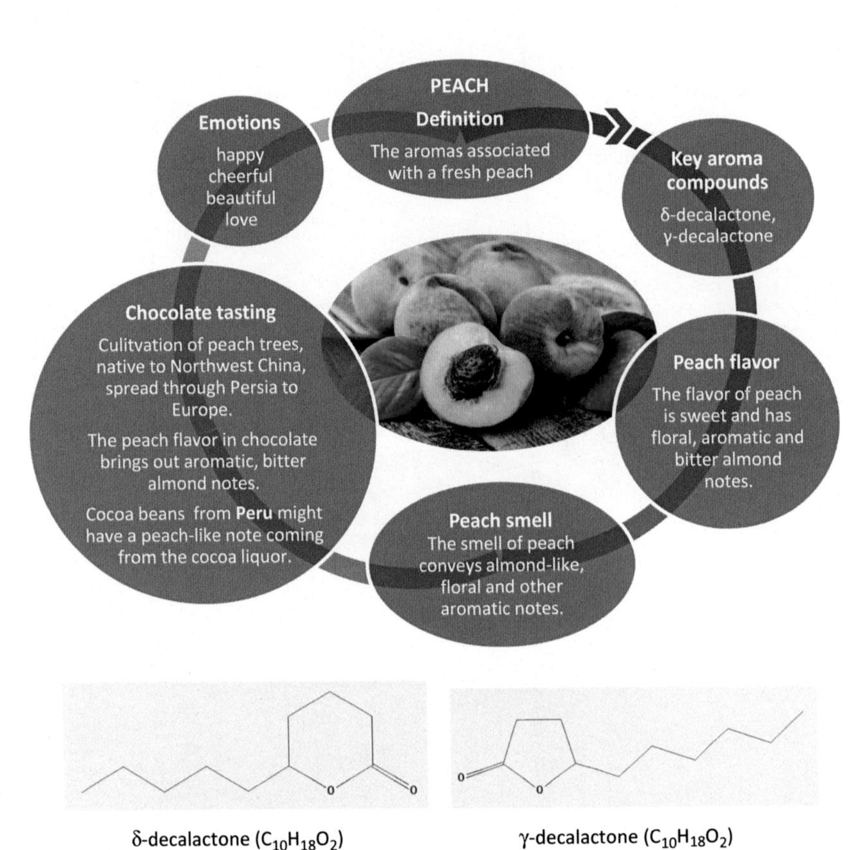

δ-decalactone ($C_{10}H_{18}O_2$)

γ-decalactone ($C_{10}H_{18}O_2$)

Apple

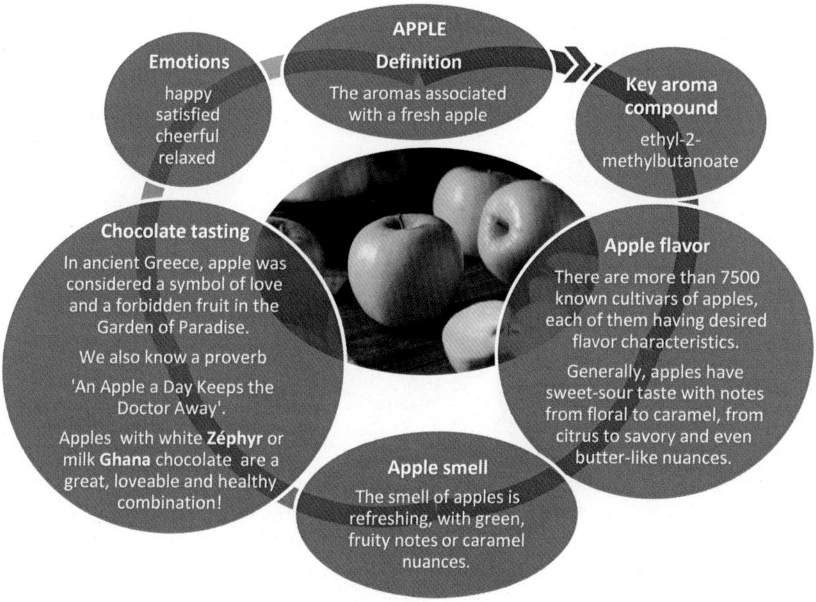

APPLE

Definition

The aromas associated with a fresh apple

Key aroma compound

ethyl-2-methylbutanoate

Apple flavor

There are more than 7500 known cultivars of apples, each of them having desired flavor characteristics.

Generally, apples have sweet-sour taste with notes from floral to caramel, from citrus to savory and even butter-like nuances.

Apple smell

The smell of apples is refreshing, with green, fruity notes or caramel nuances.

Chocolate tasting

In ancient Greece, apple was considered a symbol of love and a forbidden fruit in the Garden of Paradise.

We also know a proverb

'An Apple a Day Keeps the Doctor Away'.

Apples with white **Zéphyr** or milk **Ghana** chocolate are a great, loveable and healthy combination!

Emotions

happy
satisfied
cheerful
relaxed

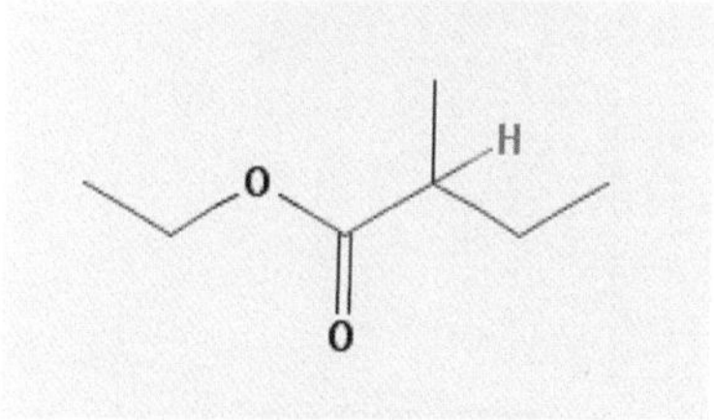

ethyl-2-methylbutanoate ($C_7H_{14}O_2$)

Pear

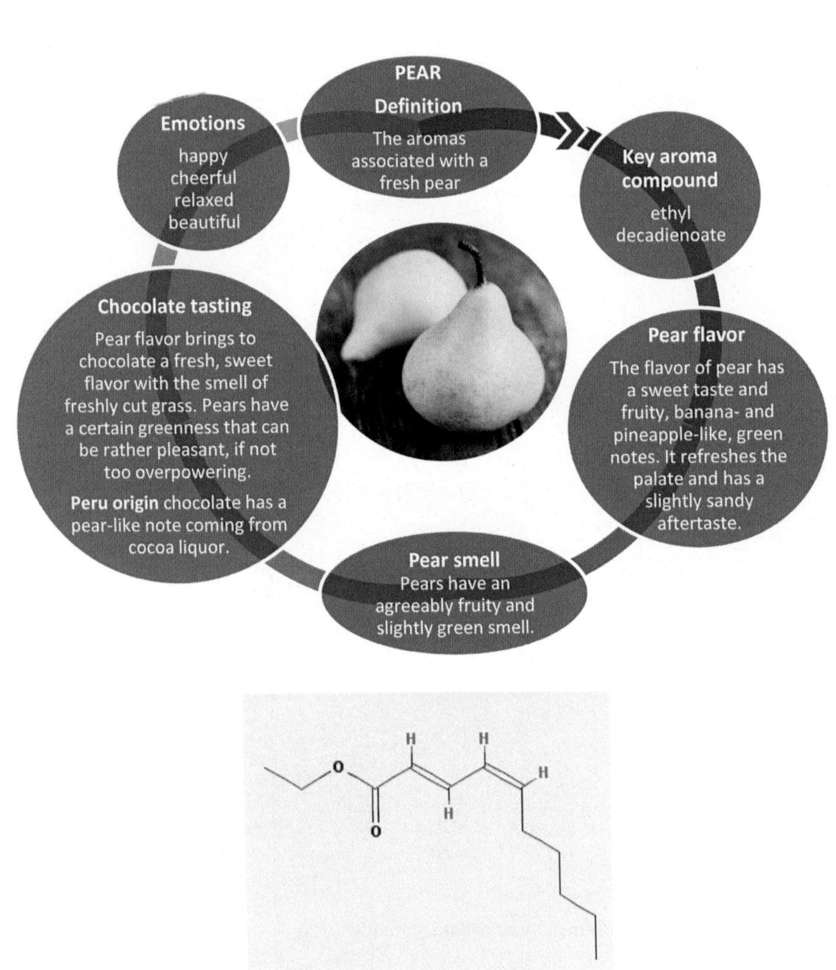

ethyl decadienoate ($C_{12}H_{20}O_2$)

Pineapple

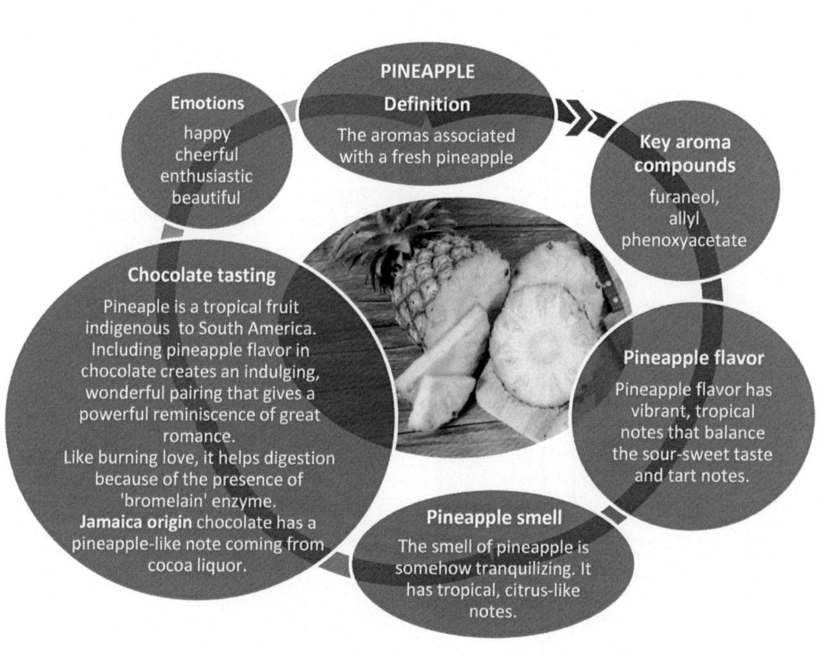

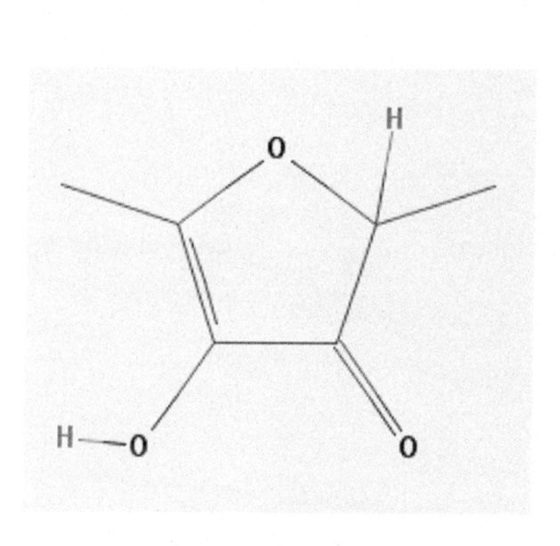

furaneol ($C_6H_8O_3$)

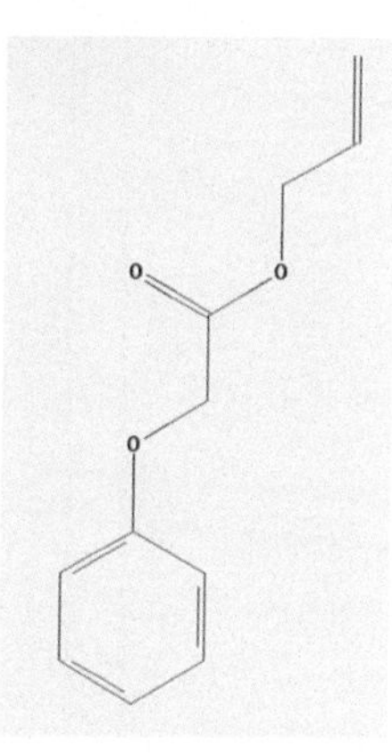

allyl phenoxyacetate ($C_{11}H_{12}O_3$)

Mango

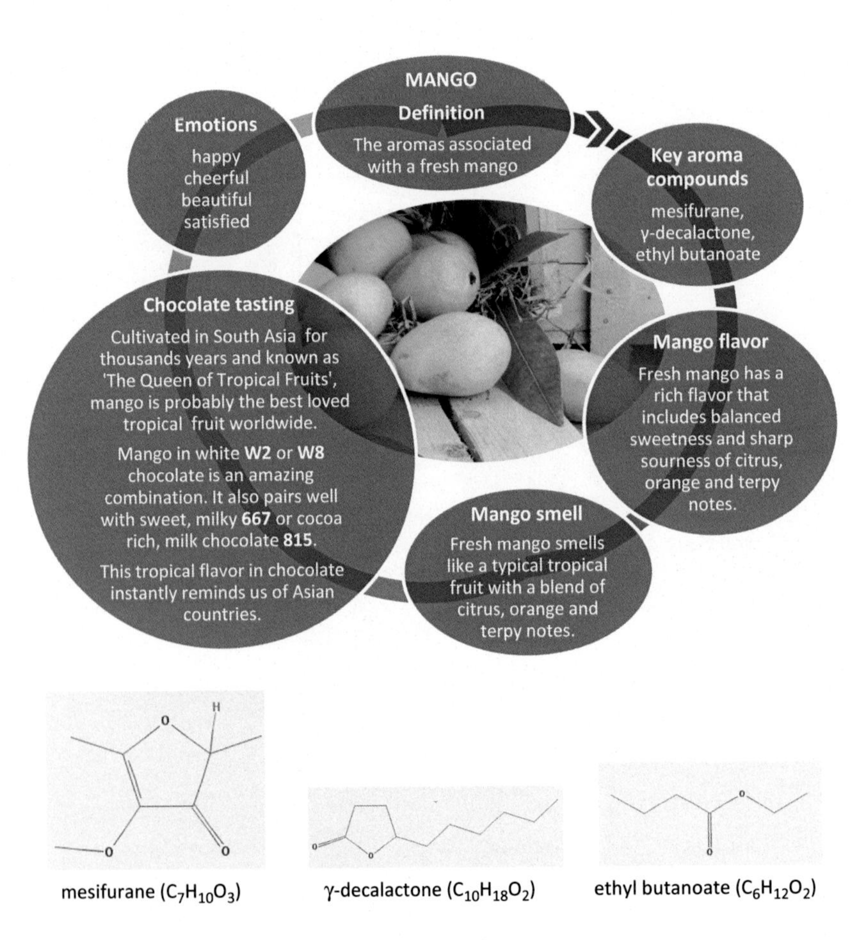

mesifurane ($C_7H_{10}O_3$)

γ-decalactone ($C_{10}H_{18}O_2$)

ethyl butanoate ($C_6H_{12}O_2$)

Banana

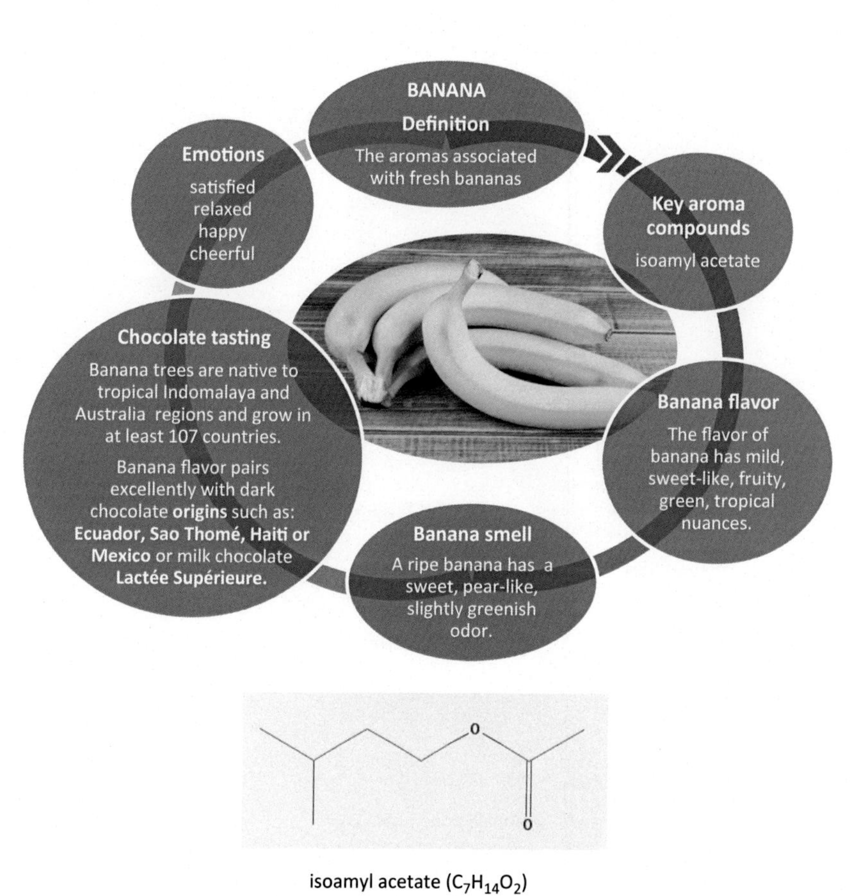

isoamyl acetate ($C_7H_{14}O_2$)

Coconut

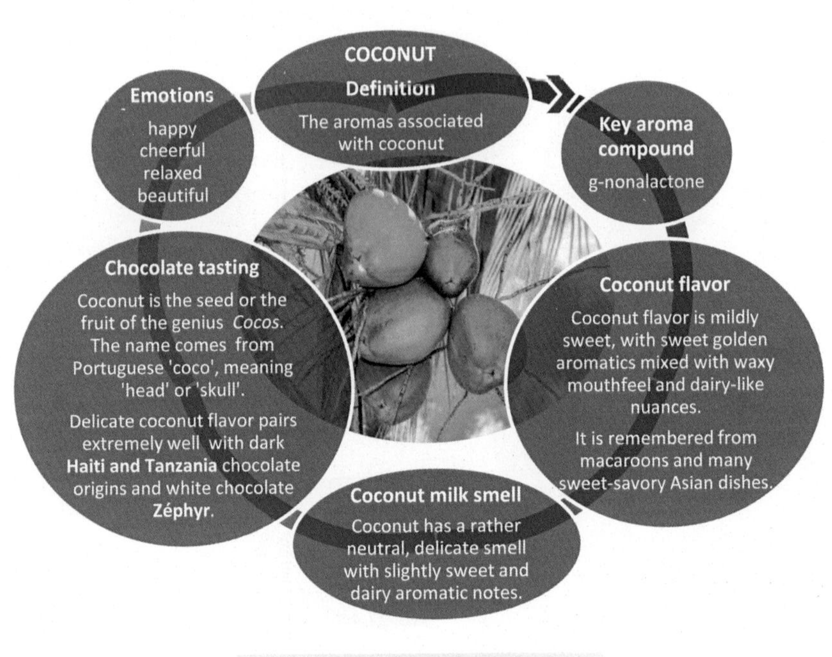

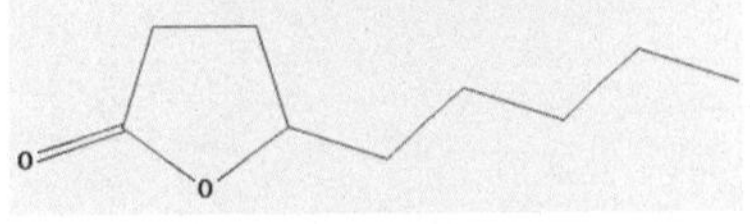

g-nonalactone ($C_9H_{16}O_2$)

RED

Grapes

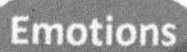

GRAPES
Definition
The aromas associated with fresh grapes

Key aroma compound
valencene

Grapes flavor
The flavor of grapes can be described as distinctively green, herbaceous and earthy. It can also be sweet, pungent and terpy.

Grapes smell
The smell of fresh grapes reminds sunny vineyards, where aromas are created. The acidic force is reduced and sugary power comes to full life.

Chocolate tasting
The combination of the unique, sweet, tart flavor of grapes and chocolate gives an impression of sweet, viney and less bitter chocolate.

Emotions
happy
cheerful
beautiful
relaxed

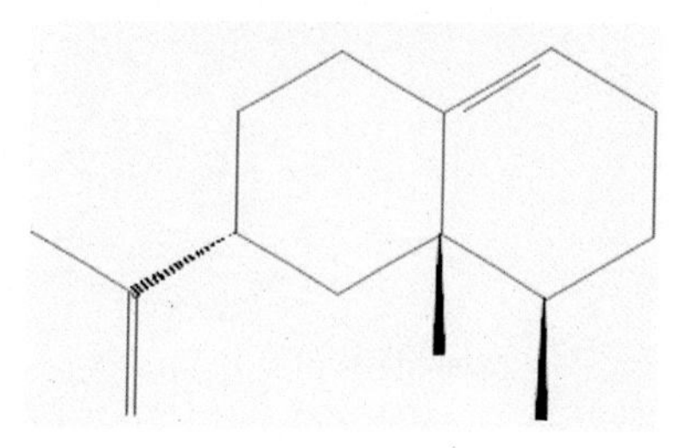

valencene ($C_{15}H_{24}$)

Cherries

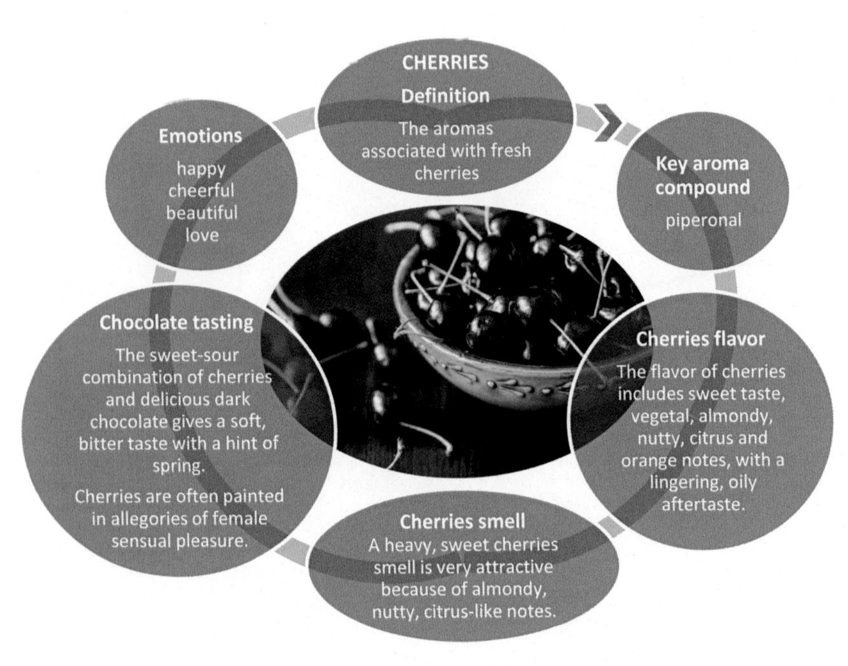

piperonal ($C_8H_6O_3$)

Raspberry

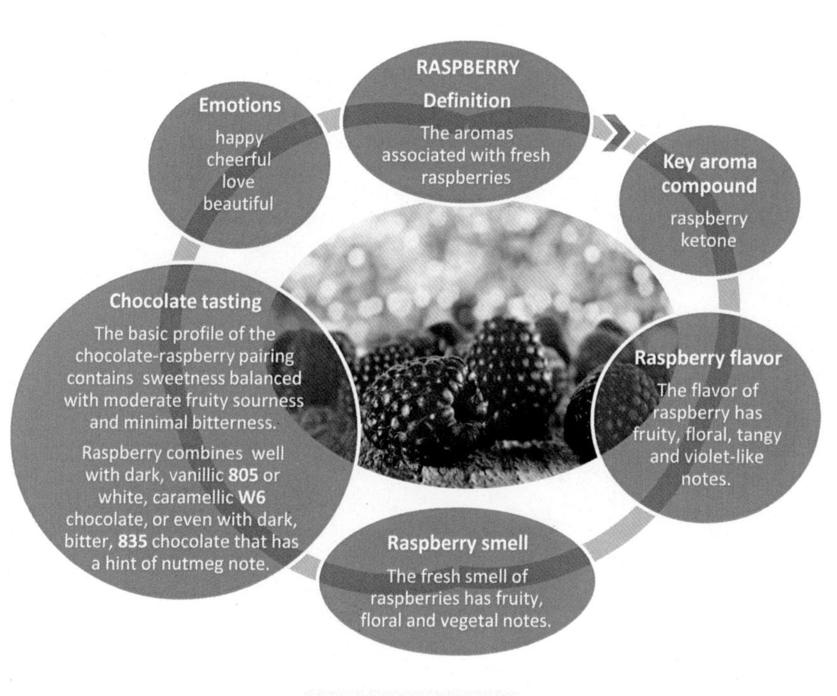

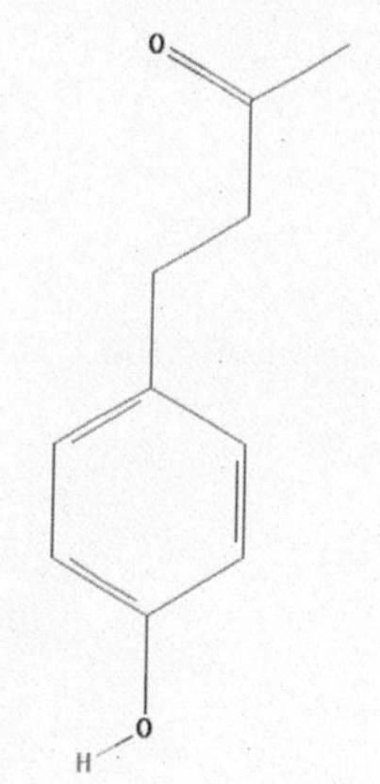

raspberry ketone ($C_{10}H_{12}O_2$)

Strawberry

STRAWBERRY
Definition
The aromas associated with strawberry

Key aroma compound
furaneol

Strawberry flavor
Strawberries have a green, waxy, sweet, buttery, fruity flavor with ripe berry nuances. Sometimes they can even have cognac-like notes.

Strawberry smell
Smell of strawberries is fresh and lifting, kind of ethereal.

Sometimes the smell includes caramel notes or 'burnt pineapple' nuances.

Chocolate tasting
With bright color, juicy texture and special sweetness, strawberries added to chocolate are a hallmark of the confectionery world.

They pair extremely well with finest Belgian **826** milk chocolate and white, very sweet **W11** chocolate.

Emotions
happy
love
beautiful
cheerful

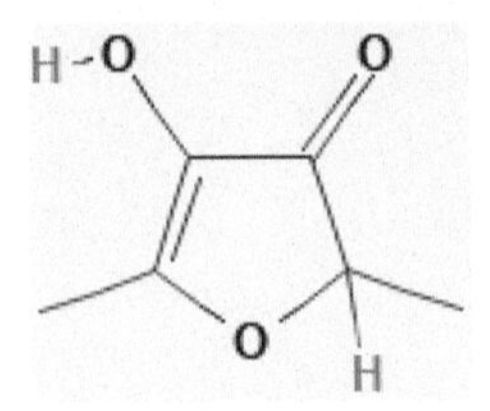

furaneol ($C_6H_8O_3$)

CITRUS

Lime

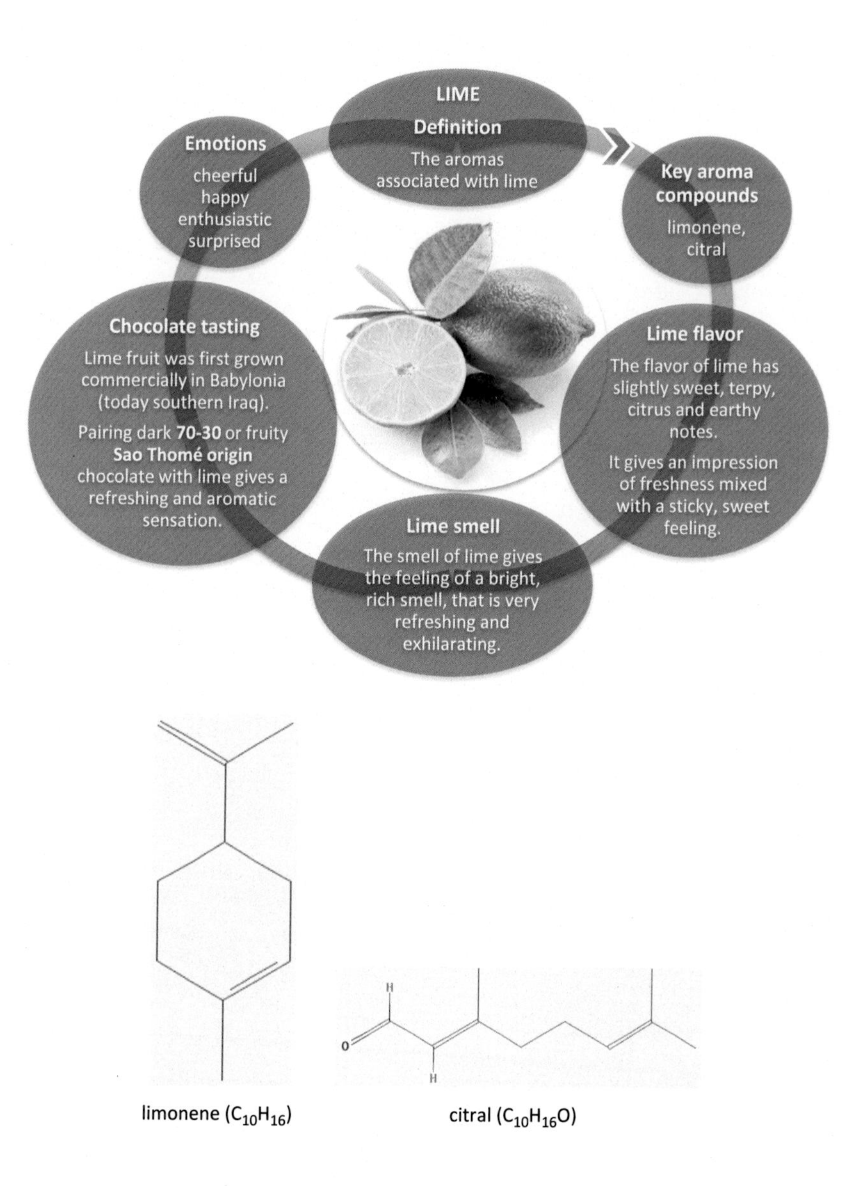

limonene ($C_{10}H_{16}$) citral ($C_{10}H_{16}O$)

Lemon

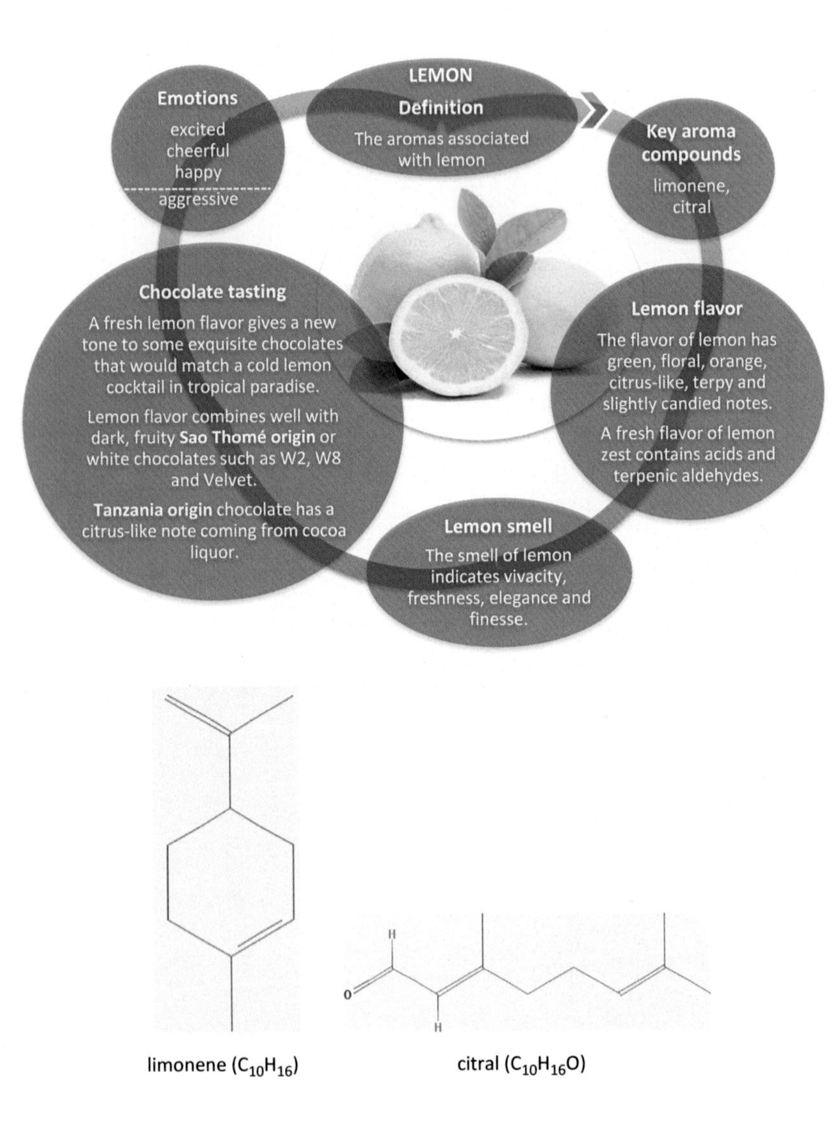

limonene ($C_{10}H_{16}$) citral ($C_{10}H_{16}O$)

Orange

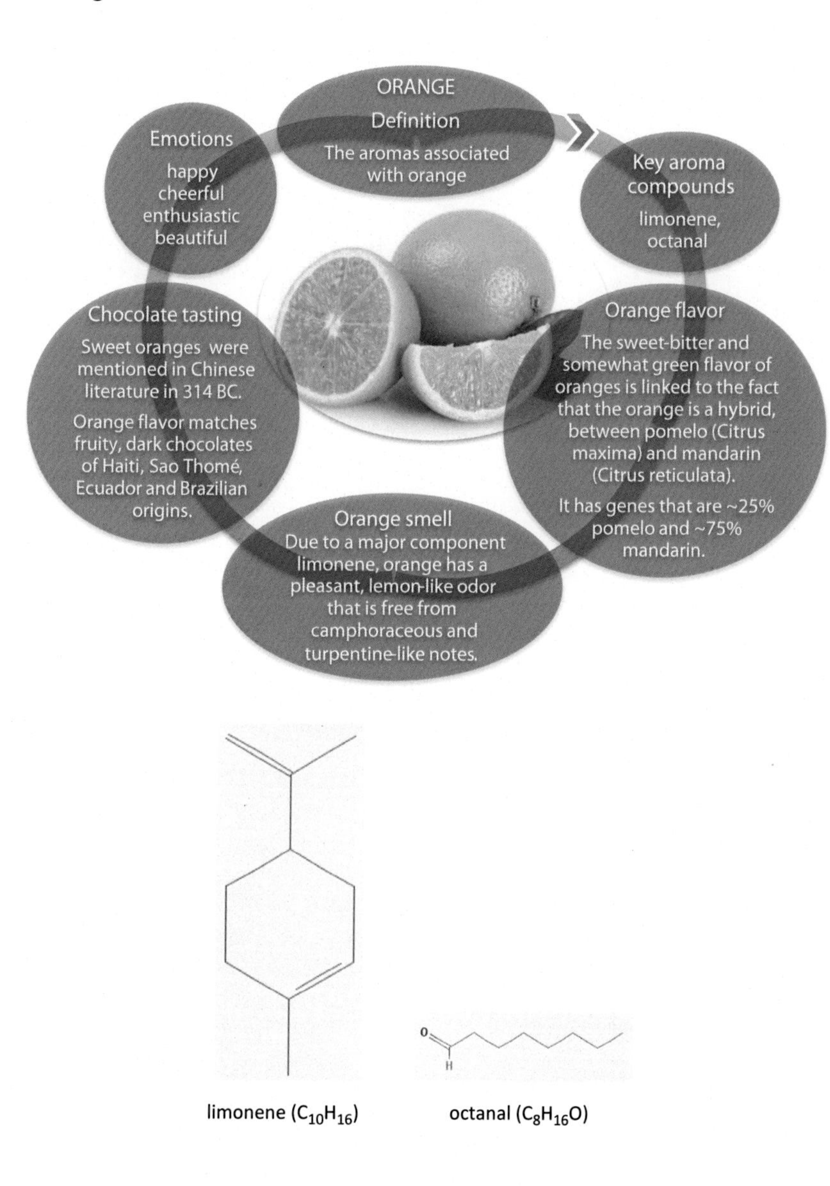

limonene ($C_{10}H_{16}$) octanal ($C_8H_{16}O$)

Mandarin

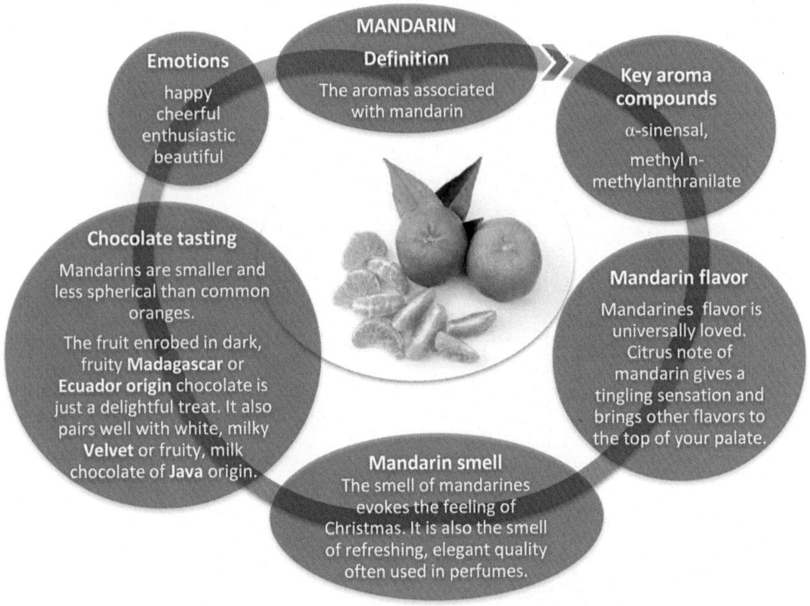

MANDARIN

Definition

The aromas associated with mandarin

Key aroma compounds

α-sinensal,

methyl n-methylanthranilate

Mandarin flavor

Mandarines flavor is universally loved. Citrus note of mandarin gives a tingling sensation and brings other flavors to the top of your palate.

Mandarin smell

The smell of mandarines evokes the feeling of Christmas. It is also the smell of refreshing, elegant quality often used in perfumes.

Chocolate tasting

Mandarins are smaller and less spherical than common oranges.

The fruit enrobed in dark, fruity **Madagascar** or **Ecuador origin** chocolate is just a delightful treat. It also pairs well with white, milky **Velvet** or fruity, milk chocolate of **Java** origin.

Emotions

happy
cheerful
enthusiastic
beautiful

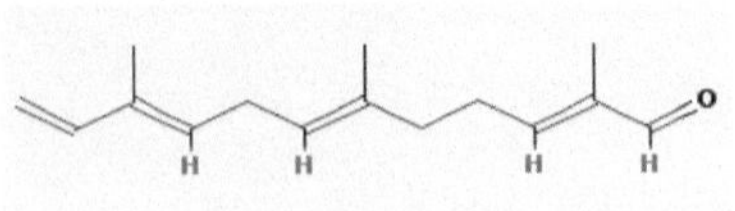

H H H H O

α-sinensal ($C_{15}H_{22}O$)

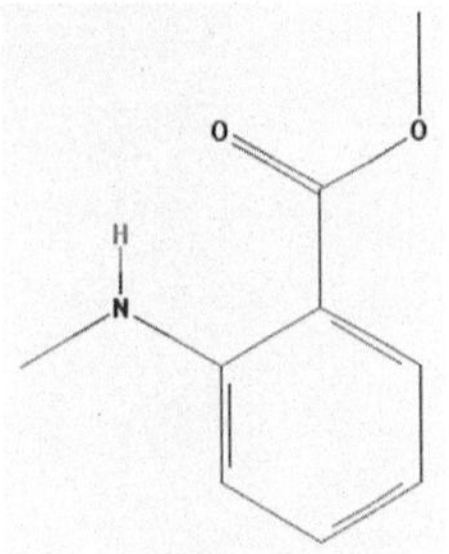

O O H N

methyl n-methylanthranilate ($C_9H_{11}NO_2$)

Tamarind

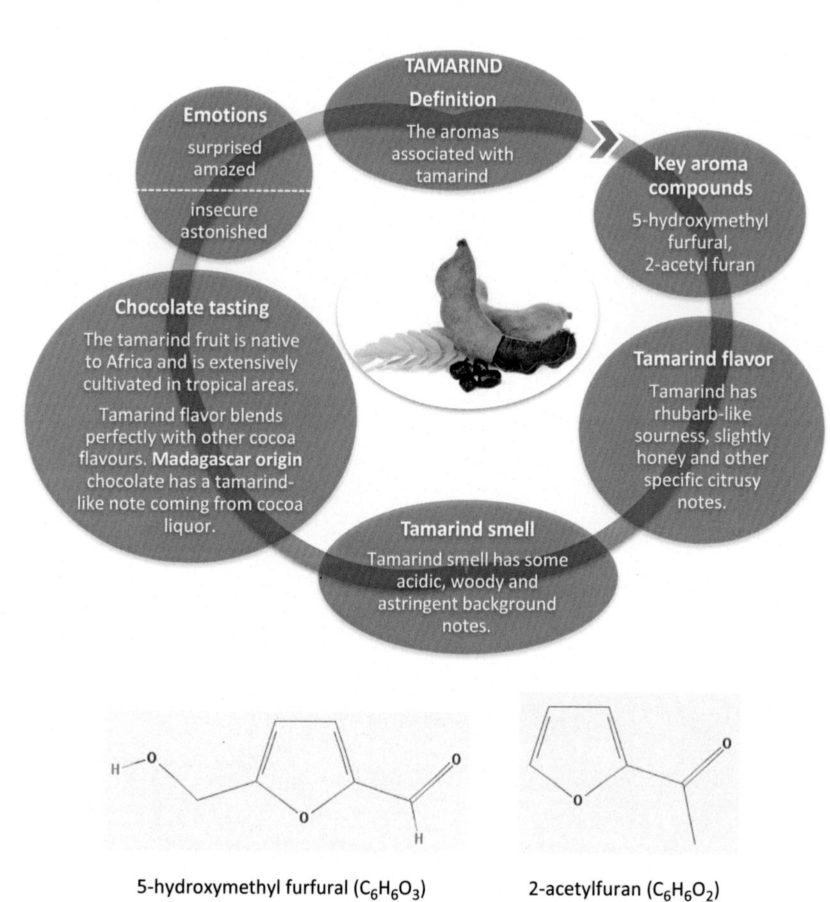

5-hydroxymethyl furfural ($C_6H_6O_3$)

2-acetylfuran ($C_6H_6O_2$)

Bergamot

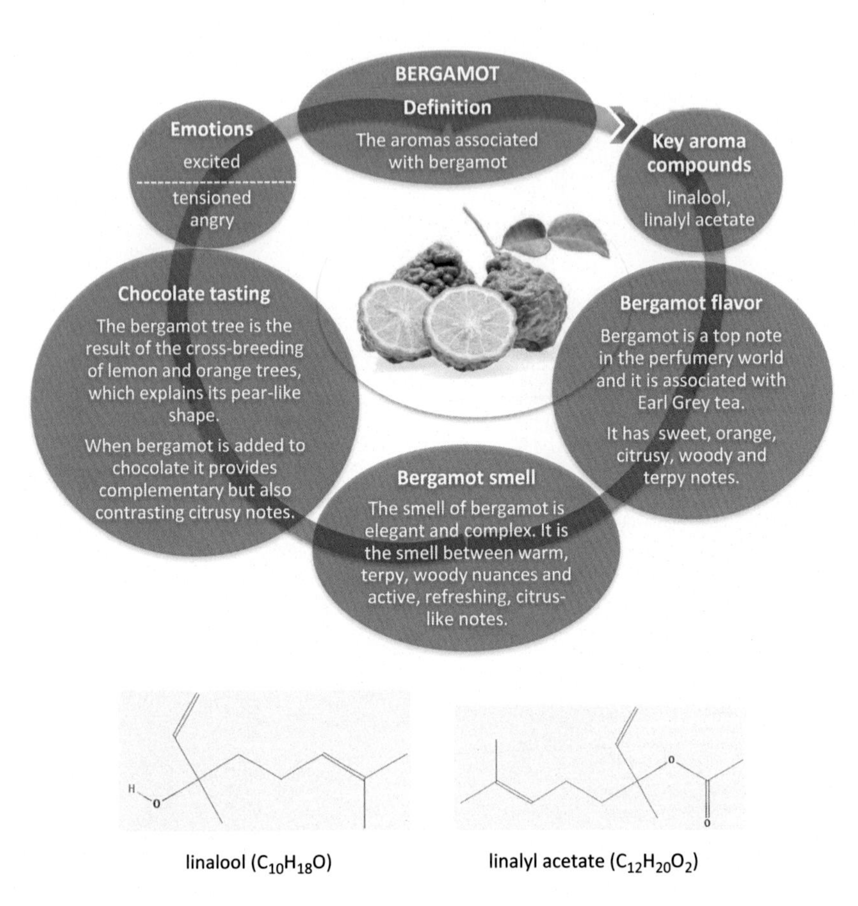

linalool ($C_{10}H_{18}O$) linalyl acetate ($C_{12}H_{20}O_2$)

FURTHER READING

Abrar, O.E.E.A., Saad, M.H.A., 2015. Chemical composition and antimalarial activity of extracts of Sudanese *Tamarindus indica* L. (Fabaceae). Pharma Innov. J. 4 (4), 90–93, (tamarind).

Akhtar, S., Mahmood, S., Naz, S., Nasir, M., Saultan, M.T., 2009. Sensory evaluation of mangoes (*Magnifera indica* L.) grown in different regions of Pakistan. Pak. J. Bot. 41 (6), 2821–2829, (mango).

Angulo, O., Fidelibus, M.W., Heymann, H., 2007. Grape cultivar and drying method affect sensory characteristics and consumer preference of raisins. J. Sci. Food Agric. 87 (5), 865–870, (raisins).

Apra, E., Biasioli, F., Gasperi, F., 2015. Volatile compounds of raspberry fruit: from analytical methods to biological role and sensory impact. Molecules 20, 2445–2474, (raspberries).

Baraem, I., Haffar, I., Baalbaki, R., Henry, J., 2001. Development of a total quality scoring system based on consumer preference weightings and sensory profiles: application to fruit dates (Tamr). Food Qual. Pref. 12, 499–506, (dates).

Belitz, H.D., Grosch, W., Schieberle, P., 2009. Food Chemistry. Springer Science & Business Media, Germany, (pineapple, raspberries).

Cheong, M.W., Liu, S.Q., Zhou, W., Curran, P., Yu, B., 2012. Chemical composition and sensory profile of pomelo (*Citrus grandis* (L.) Osbeck) juice. Food Chem. 135 (4), 2505–2513, (lemon).

Dever, M.C., MacDonald, R.A., Cliff, M.A., Lane, W.D., 1996. Sensory evaluation of sweet cherry cultivars. Hortscience 31 (1), 150–153, (cherries).

Didzbalis, J., Munafo, J.P., 2014. Mango Flavor Compositions. , (mango).

Galmarini, M.V., Zamora, M.C., Baby, R., Chirife, J., Mesina, V., 2008. Aromatic profiles of spray-dried encapsulated orange flavors: influence of matrix composition on the aroma retention evaluated by sensory analysis and electronic nose techniques. Int. J. Food Sci. Technol. 43 (9), 1569–1576, (mandarin).

Genova, G., Montanaro, G., Perkin, E., 2012. Qualitative Evaluation of Aroma-Active Compounds in Grape and Grape-Derived Products by Means of Headspace SPME-GC/MS Analysis, pp. 1–7. (raisins).

Getchell, T.V., 1988. Molecular, Membranous, and Cytological Studies. F.L. Margolis, New Jersey, (cherries).

Greger, V., Schieberle, P., 2007. Characterization of the key aroma compounds in apricots (*Prunus armeniaca*) by application of the molecular sensory science concept. J. Agric. Food Chem. 55 (13), 5221–5228, (apricots).

Guichard, E., Schlich, P., Issanchou, S., 1990. Composition of apricot aroma: correlations between sensory and instrumental data. J. Food Sci. 55 (3), 735–738, (apricot).

Haug, M.T., King, E.S., Heymann, H., Crisosto, C.H., 2003. Sensory profiles for dried fig (*Ficus carica* L.) cultivars commercially grown and processed in California. J. Food Sci. 78 (8), 1273–1281, (figs).

http://www.bojensen.net/EssentialOilsEng/EssentialOils.htm. (grapes, lemon, lime, mandarin, bergamot).

Jaeger, S.R., Lund, C.M., Lau, K., Harker, F.R., 2003. In search of the "Ideal" pear (*Pyrus* spp.): results of a multidisciplinary exploration. J. Food Sci. 68 (3), 1108–1117, (pear).

King, E.S., Hopfer, H., Haug, M.T., Orsi, J.D., Heymann, H., Crisosto, G.M., Crisosto, C.H., 2012. Describing the appearance and flavor profiles of fresh fig (*Ficus carica* L.) cultivars. J. Food Sci. 77 (12), S419–S429, (figs).

Lasekan, O., See Ng, S., 2015. Key volatile aroma compounds of three black velvet tamarind (Dialium) fruit species. Food Chem. 168, 561–565, (tamarind).

Le Moignea, M., Maurya, C., Bertrand, D., Jourjona, F., 2008. Sensory and instrumental characterisation of Cabernet Franc grapes according to ripening stages and growing location. Food Qual. Prefer. 19 (2), 220–231, (grapes).

Lin, J.C.C., Nagy, S., Klim, M., 1993. Application of pattern recognition techniques to sensory and gas chromatographic flavor profiles of natural orange aroma. Food Chem. 47 (3), 235–245, (mandarin).

Munafo, J.P., Didzbalis, J., Schnell, R.J., Steinhaus, M., 2016. Insights into the key aroma compounds in mango (*Mangifera indica* L. 'Haden') fruits by stable isotope dilution quantitation and aroma simulation experiments. J. Agric. Food Chem. 64 (21), 4312–4318, (mango).

Nabiha, B., Abdelfatteh, E.O., Faten, K., Hervé, C., Moncef, C.M., 2010. Chemical composition of bergamot (Citrus Bergamia Risso) essential oil obtained by hydrodistillation. J. Chem. Chem. Eng. 4 (4), 60–62, (Serial No. 29), (bergamot).

Pickenhagen, W., Velluz, A., Passerat, J.-P., Ohloff, G., 1981. Estimation of 2,5-dimethyl-4-hydroxy-3(2H)-furanone (FURANEOLA®) in cultivated and wild strawberries, pineapples and mangoes. J. Sci. Food Agric. 32 (11), 1132–1134, (strawberries).

Pilar Cano, M., de Ancos, B., Cruz Montallana, M., Camara, M., Reglero, G., Tabera, J., 1997. Differences among Spanish and Latin-American banana cultivars: morphological, chemical and sensory characteristics. Food Chem. 59 (3), 411–419, (banana).

Pons, A., Lavigne, V., Frérot, E., Darriet, P., Dubourdieu, D., 2008. Identification of volatile compounds responsible for prune aroma in prematurely aged red wines. J. Agric. Food Chem. 56 (13), 5285–5290, (prunes).

Predieri, S., Ragazzini, P., Rondelli, R., 2006. Sensory evaluation and peach fruit quality. Acta Hortic. 713, 429–434, (pineapple).

Rétiveau, A.N., Chambers, IV.E., Millken, G.A., 2004. Common and specific effects of fine fragrances on the mood of women. J. Sens. Stud. 19, 373–394, (citrus).

Rothe, M., 1988. Introduction to Aroma Research. Kluwer Academic Publishers, London, (pear).

Sabarez, H.T., Price, W.E., Korth, J., 2000. Volatile changes during dehydration of d'Agen prunes. J. Agric. Food Chem. 48 (5), 1838–1842, (prunes).

Song, J., Bangerth, F., 1996. The effect of harvest date on aroma compound production from 'Golden Delicious' apple fruit and relationship to respiration and ethylene production. Postharvest Biol. Technol. 8 (4), 259–269, (apple).

Takeoka, G.R., Flath, R.A., Mon, T.R., Teranishi, R., Guentert, M., 1990. Volatile constituents of apricots (*Prunus armeniaca*). J. Agric. Food Chem. 38, 471–477, (apricots).

Tokitomo, Y., Steinhaus, M., Buttner, A., Schieberle, P., 2005. Odor-active constituents in fresh pineapple (*Ananas comosus* [L.] Merr.) by quantitative and sensory evaluation. Biosci. Biotechnol. Biochem. 69 (7), 1323–1330, (pineapple).

Xu, Q., Chen, L.-L., Ruan, X., Chen, D., Zhu, A., Chen, C., Bertrand, D., Jiao, W.-B., Hao, B.-H., Lyon, M.P., Chen, J., Gao, S., Xing, F., Lan, H., Chang, J.-W., Ge, X., Lei, Y., Hu, Q., Miao, Y., Wang, L., Xiao, S., Biswas, M.K., Zeng, W., Guo, F., Cao, H., Yang, X., Xu, X.-W., Cheng, Y.-J., Xu, J., Liu, J.-H., Luo, O.J., Tang, Z., Guo, W.-W., Kuang, H., Zhang, H.-Y., Roose, M.L., Nagarajan, N., Deng, X.-X., Ruan, Y., 2013. The draft genome of sweet orange (*Citrus sinensis*). Nat. Genet. 45, 59–66, (orange).

Chapter 6

Botanical

Chapter Outline

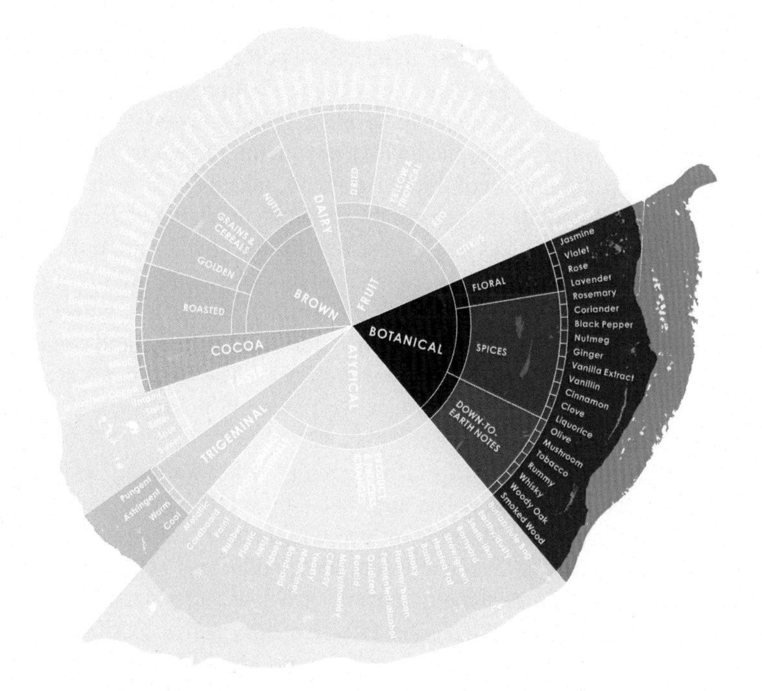

Hidden Persuaders in Cocoa and Chocolate. http://dx.doi.org/10.1016/B978-0-12-815447-2.00007-3

Botanical notes in cocoa and chocolate refer to those impressions that remind us about closeness and richness of nature. They also represent a kind of changing 'Autumn and Winter' atmosphere. They certainly add to the complexity of the overall flavor of cocoa liquors of certain origins and are subtly present in chocolate made from these liquors. These impressions may also be called 'herbaceous' or 'vegetal' and may be linked to notes detected in unripe, raw or spicy products.

In this chapter, we describe the three groups of botanical notes: *floral, spicy* and so called *'down-to-earth'*. They are even more rare and precious than fruity ones. They give a special meaning in high quality, 'gourmet' products. They are clearly linked to the cocoa origin and botanical variety of the cocoa trees. As they are found in some exclusive raw material, they can be truly linked to a specific cocoa plantation and sometimes even to a specific fermentation process.

Firstly, 'floral' notes may be quite volatile and linked to *jasmine, lavender, rose and violet*; the flowers used as fragrance ingredients in the perfumery sector. It is well known in the fragrance industry that some of them may be on a more singular and active side, like lavender or on a more complex and warm sedative side as jasmine or rose.

Secondly, 'spicy' notes are mostly warm and stimulating. Hereby, we focus on *vanilla extract, vanillin, clove, cinnamon, nutmeg, ginger, pepper, liquorice, rosemary, coriander.* Some of them are often associated with 'brown spices' mostly used during Christmas period or in specific, known brown cookies called speculaas, or brownies.

Thirdly, 'down-to-earth' notes include those perceptions that are frequently seen as representing forest or quite 'raw' natural impressions, such as *olive, tobacco, whisky, rummy, mushroom, woody, smoked wood.* Many of these notes would be certainly associated with a cozy winter evening when you tuck yourself in a warm blanket, put the fire glowing, take a sip of wine or whisky, and enjoy a nice chocolate dessert combined with surprising, delicious spices.

The majority of botanical notes presented in this chapter are critically important in the perfumery sector that classifies them basically into four corners of the famous fragrance wheel of Edwards from 1993. In this wheel we can distinguish two parts of rather 'warm' notes: floral and oriental, and two parts of rather 'cold' notes, including fresh and woody ones. The fresh notes are further on distinguished into green and citrusy, while woody notes are split into dry wood, mossy wood and woody oriental notes.

One might think that the aspect of 'cooling' and 'warming' of botanical notes is less important in the cocoa and chocolate sector. On the contrary, as a part of the creative process, developers are continuously pairing chocolate with unlimited ingredients that may also bring these additional sensory sensations. To please the ever more demanding customer, a sense of play, originality and rarity is a must over the classical, cost effective, easy and accessible chocolate approach.

Finally, regarding the link between emotions and botanical notes, we have noticed that there could be a strong polarization between what people judge as bringing positive or negative sensory impression. Like with fragrances, some people will love a touch of a certain botanical note in cocoa or chocolate, while others will actually not like it. Further research in that field is necessary because the overall sensory and emotional response may also be highly influenced by all kinds of extrinsic factors, including: the packaging, presence or lack of the information associated with the product, the occasion or time of consumption, etc. We shall also not forget the deep-seated childhood memories that some of us nature and consider as the benchmark of these unusual notes.

FLORAL

Jasmine

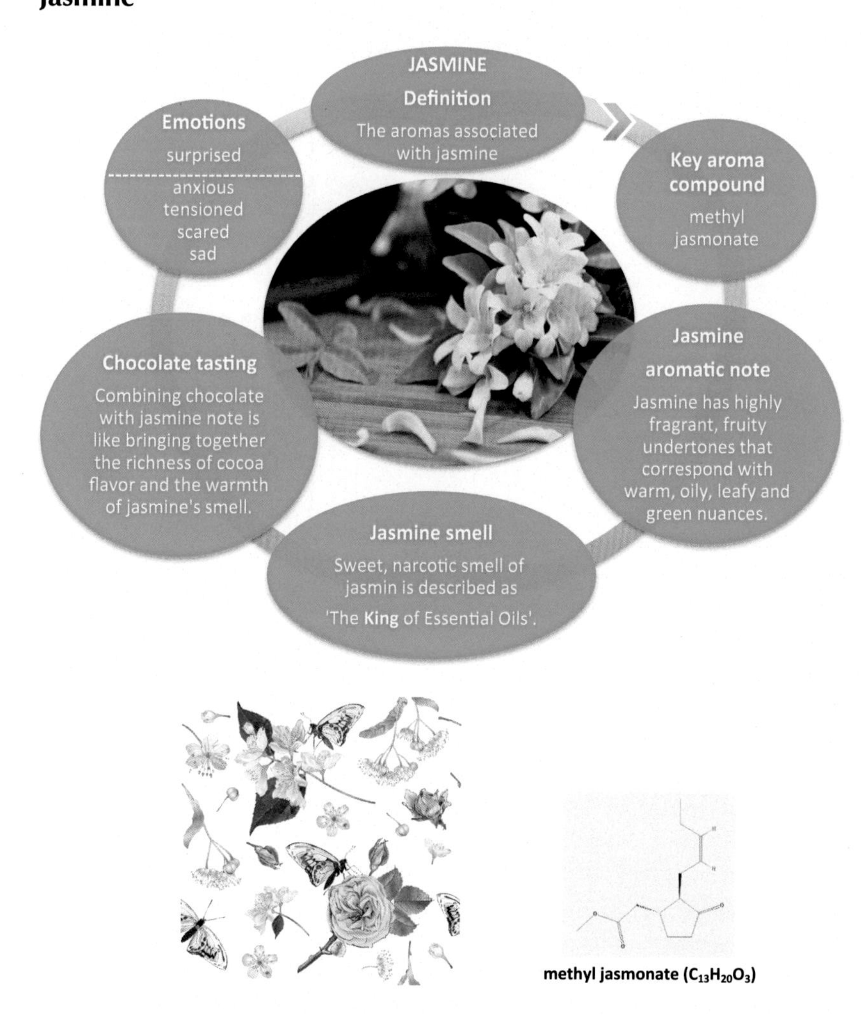

methyl jasmonate ($C_{13}H_{20}O_3$)

Violet

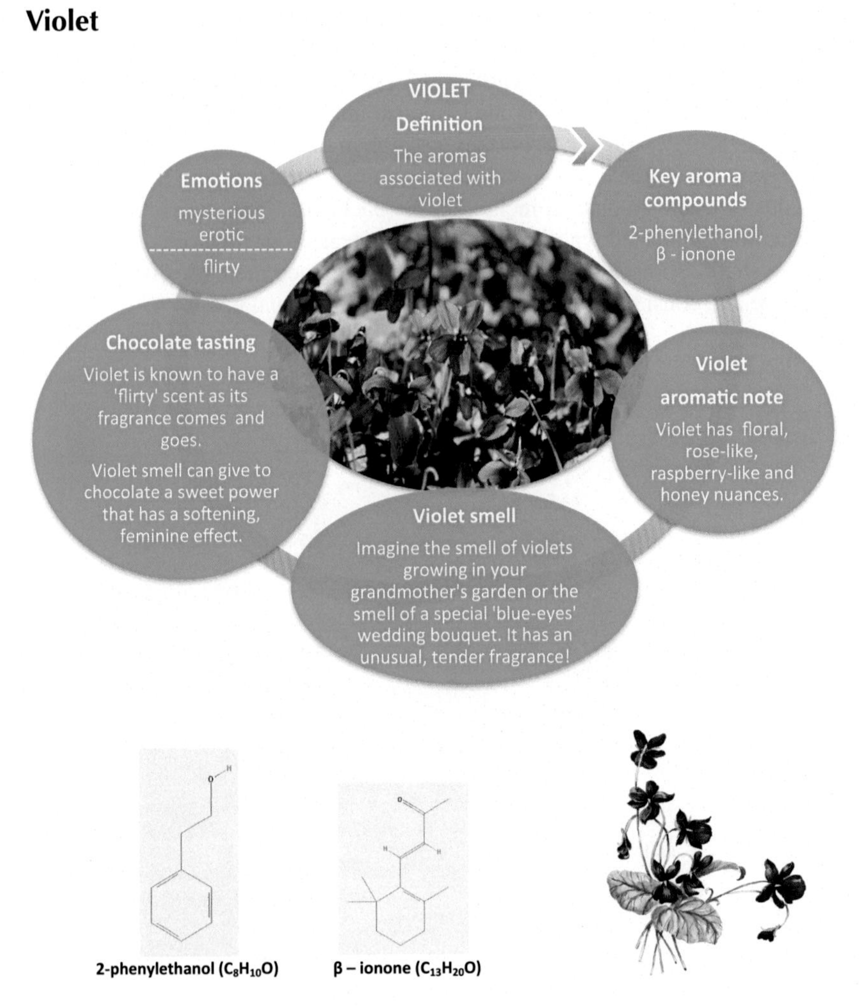
VIOLET
Definition
The aromas associated with violet
Key aroma compounds
2-phenylethanol, β - ionone
Violet aromatic note
Violet has floral, rose-like, raspberry-like and honey nuances.
Violet smell
Imagine the smell of violets growing in your grandmother's garden or the smell of a special 'blue-eyes' wedding bouquet. It has an unusual, tender fragrance!
Chocolate tasting
Violet is known to have a 'flirty' scent as its fragrance comes and goes.
Violet smell can give to chocolate a sweet power that has a softening, feminine effect.
Emotions
mysterious
erotic
flirty
2-phenylethanol ($C_8H_{10}O$)
β – ionone ($C_{13}H_{20}O$)

Rose

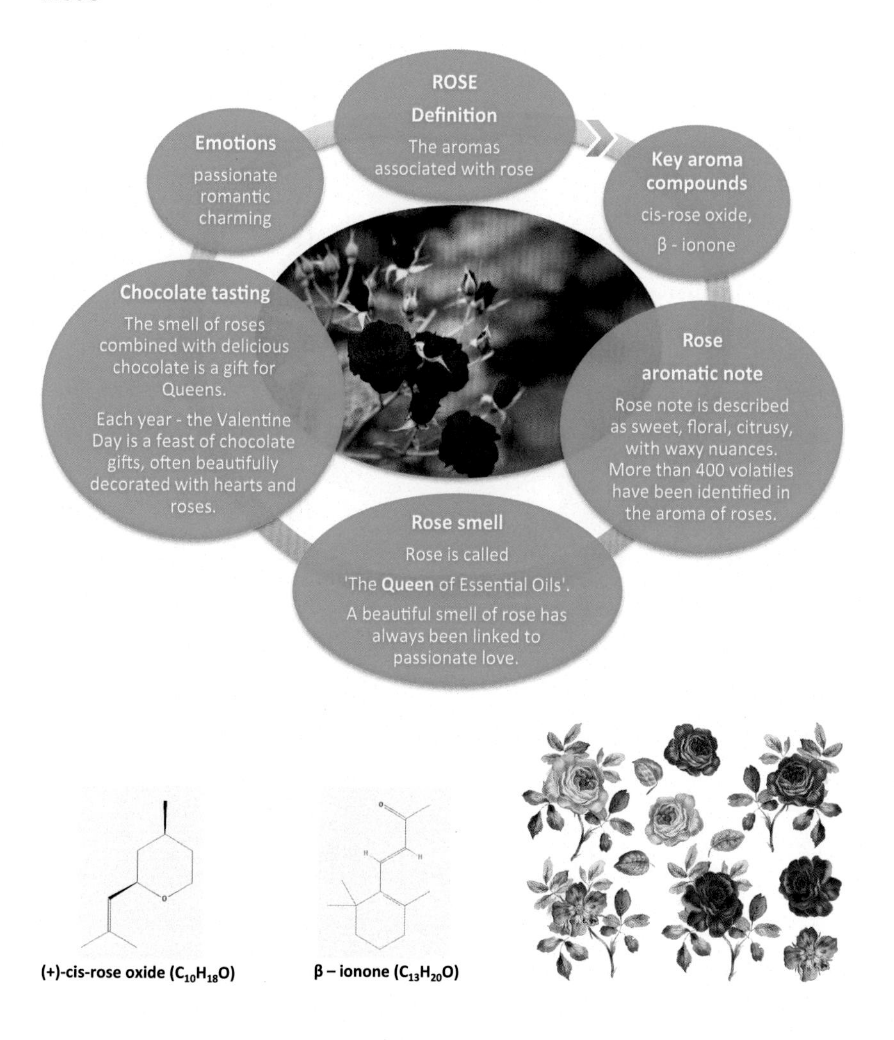
ROSE
Definition
The aromas associated with rose
Key aroma compounds
cis-rose oxide,
β - ionone
Rose
aromatic note
Rose note is described as sweet, floral, citrusy, with waxy nuances. More than 400 volatiles have been identified in the aroma of roses.
Rose smell
Rose is called
'The Queen of Essential Oils'.
A beautiful smell of rose has always been linked to passionate love.
Chocolate tasting
The smell of roses combined with delicious chocolate is a gift for Queens.
Each year - the Valentine Day is a feast of chocolate gifts, often beautifully decorated with hearts and roses.
Emotions
passionate
romantic
charming
(+)-cis-rose oxide ($C_{10}H_{18}O$)
β – ionone ($C_{13}H_{20}O$)

Lavender

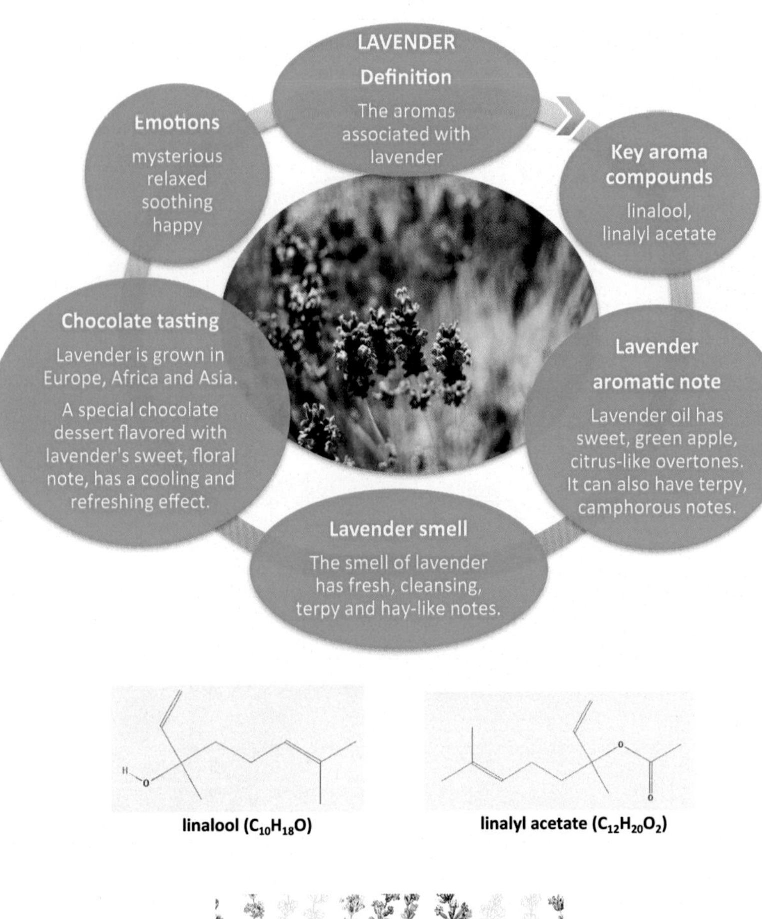
LAVENDER
Definition
The aromas associated with lavender
Key aroma compounds
linalool, linalyl acetate
Lavender aromatic note
Lavender oil has sweet, green apple, citrus-like overtones. It can also have terpy, camphorous notes.
Lavender smell
The smell of lavender has fresh, cleansing, terpy and hay-like notes.
Chocolate tasting
Lavender is grown in Europe, Africa and Asia.
A special chocolate dessert flavored with lavender's sweet, floral note, has a cooling and refreshing effect.
Emotions
mysterious
relaxed
soothing
happy
linalool ($C_{10}H_{18}O$)
linalyl acetate ($C_{12}H_{20}O_2$)

SPICES

Rosemary

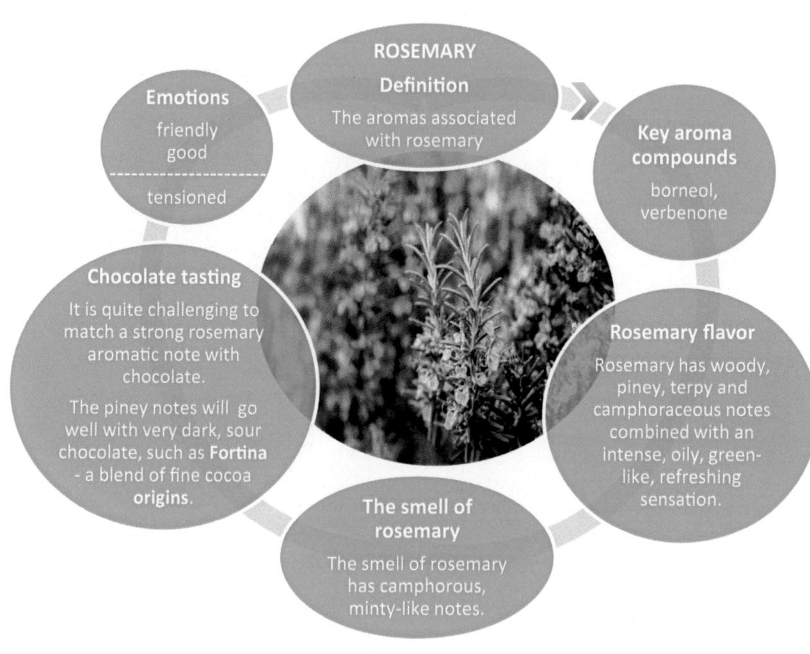

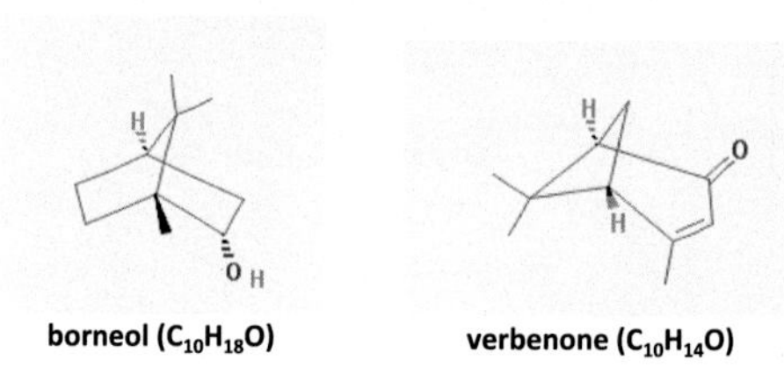

borneol ($C_{10}H_{18}O$) **verbenone ($C_{10}H_{14}O$)**

Coriander

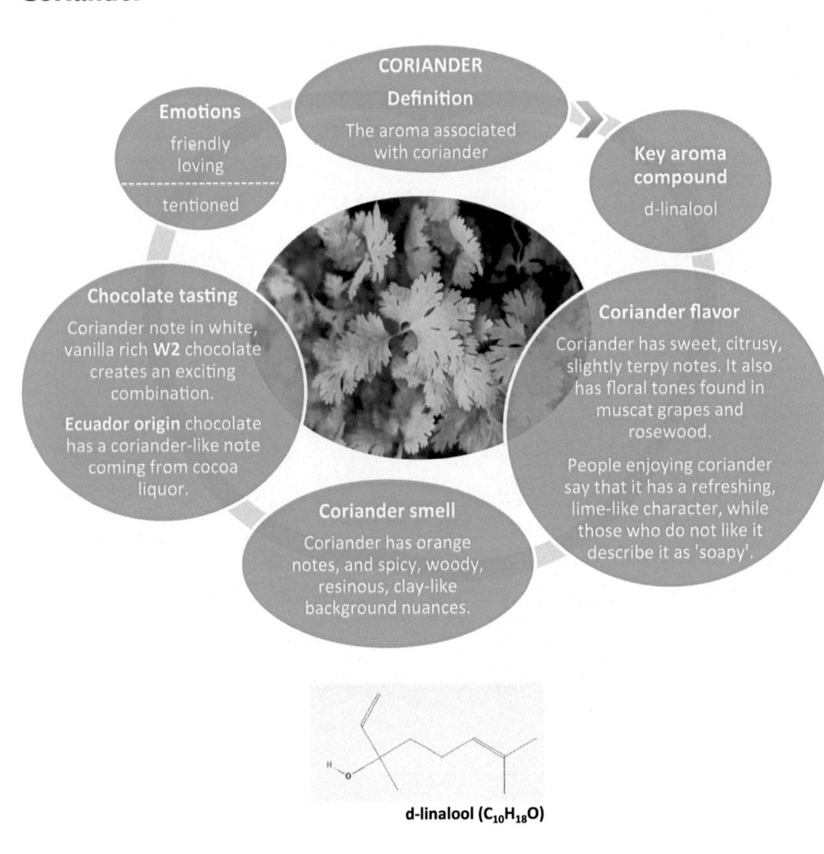

d-linalool ($C_{10}H_{18}O$)

Black pepper

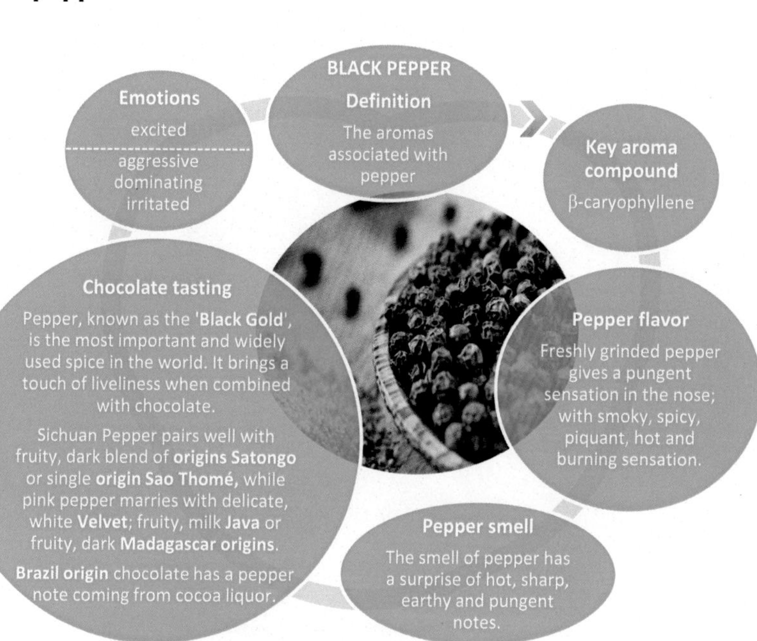

Foliage and fruit of the pepper plant

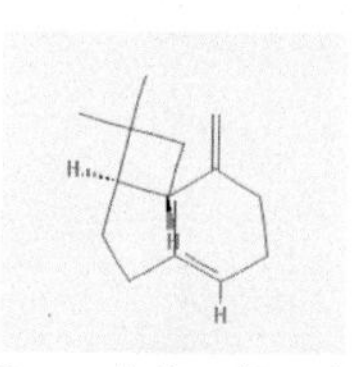

β-caryophyllene ($C_{15}H_{24}$)

Nutmeg

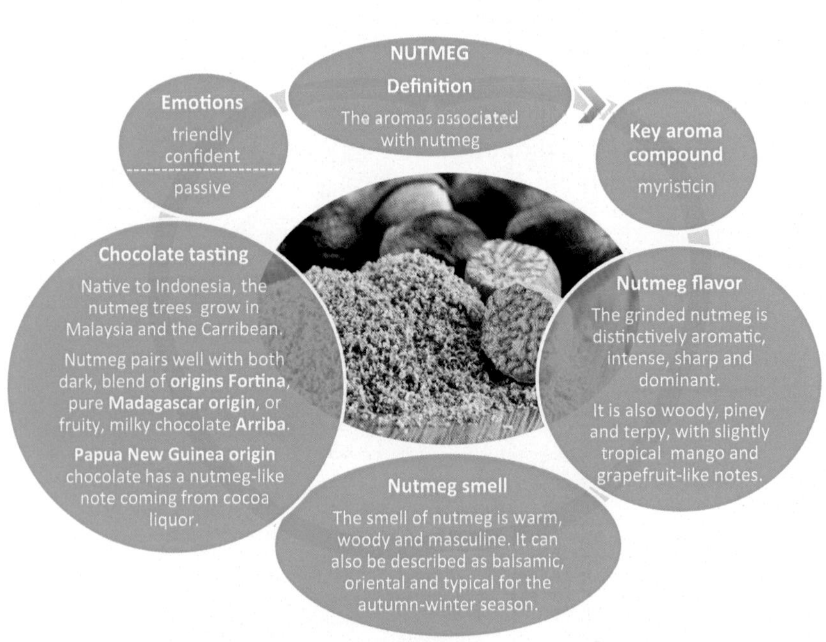

Nutmeg fruit

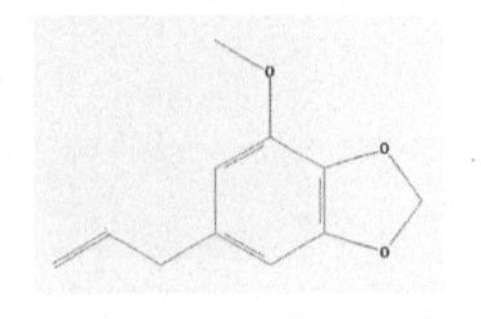

myristicin ($C_{11}H_{12}O_3$)

Ginger

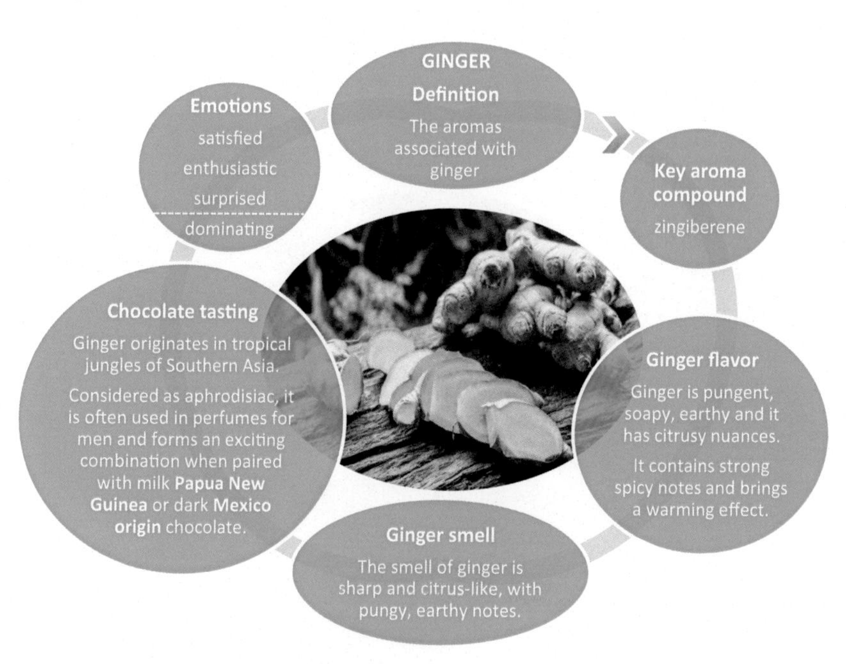

Ginger treat

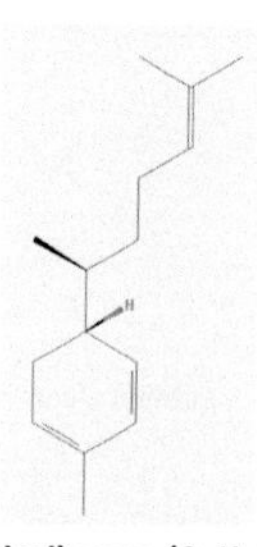

zingiberene ($C_{15}H_{24}$)

Vanilla extract

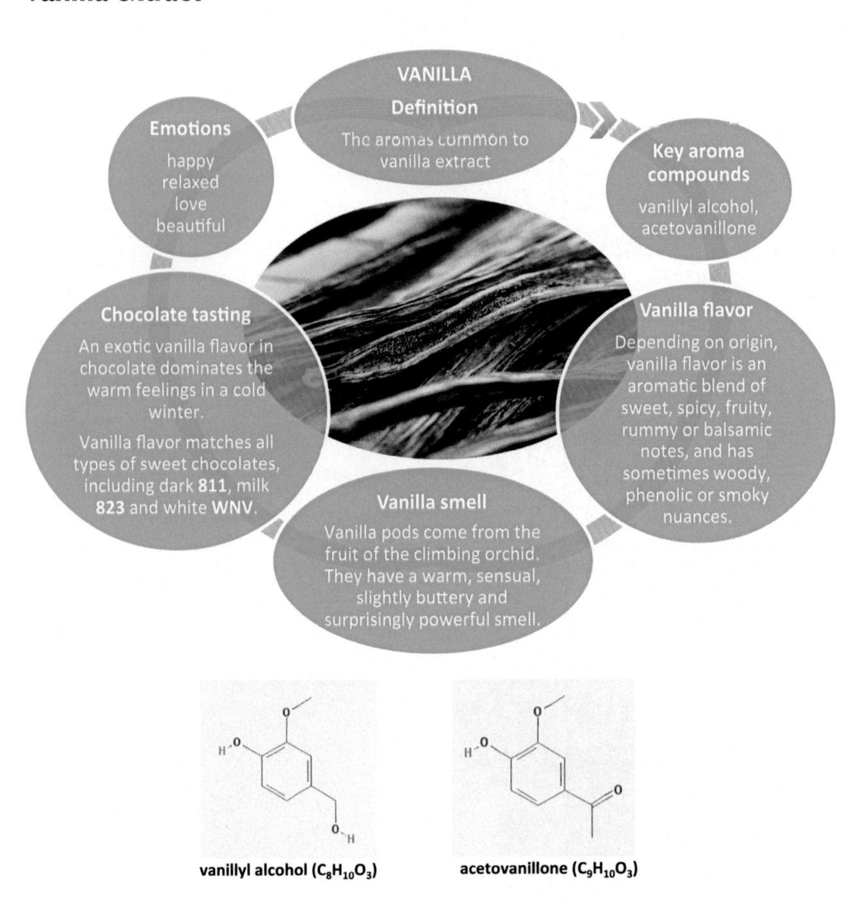

vanillyl alcohol ($C_8H_{10}O_3$)

acetovanillone ($C_9H_{10}O_3$)

Vanillin

VANILLIN

Definition

The aromas common to synthetic vanillin

Key aroma compounds

4-methylguaiacol, vanillin

Vanillin flavor

Vanillin flavor is more round and uniform than natural vanilla extract flavor.

It is sweeter and sometimes phenolic due to the vanillin molecule and number of other chemical compounds.

Vanillin smell

The vanillin smell is sweeter and sharper than vanilla smell and reminds of homemade cake.

Chocolate tasting

The effect of vanillin in chocolate is similar to vanilla extract flavor. All types of chocolates need a touch of good vanilla or vanillin.

Emotions

happy
relaxed
love
beautiful

4-methylguaiacol ($C_8H_{10}O_2$)

vanillin ($C_8H_8O_3$)

Cinnamon

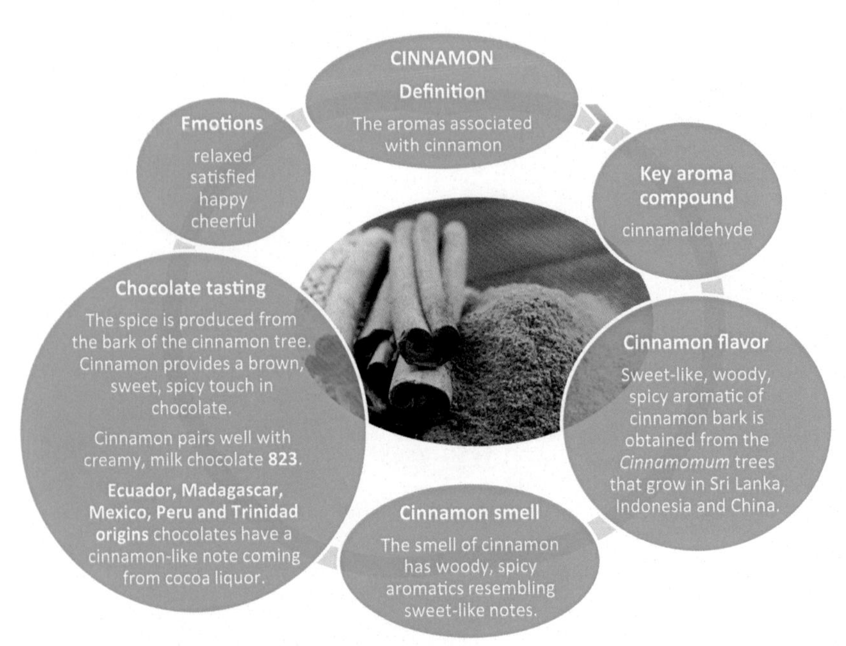

Cinnamon delight

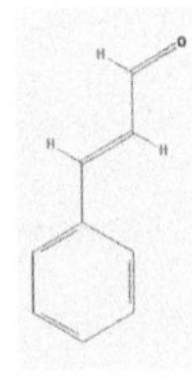

cinnamaldehyde (C_9H_8O)

Clove

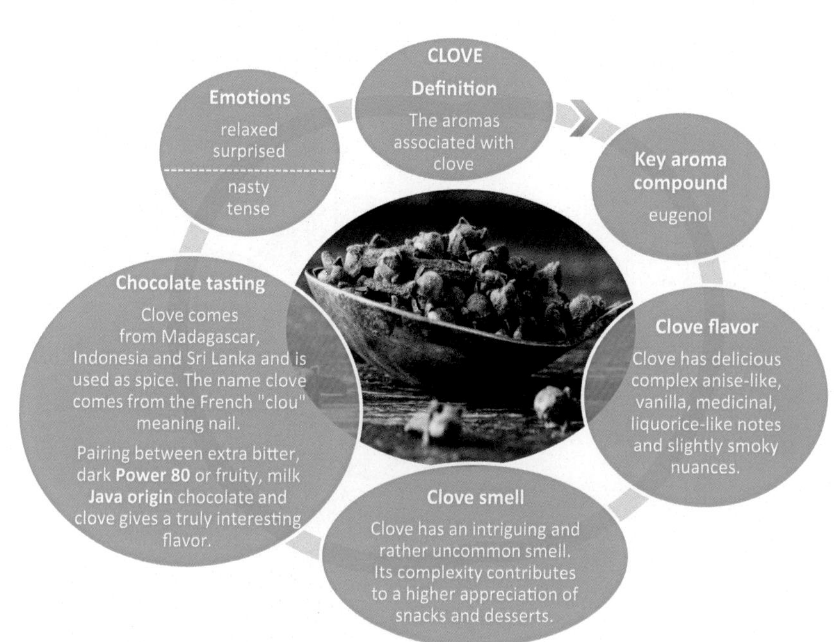

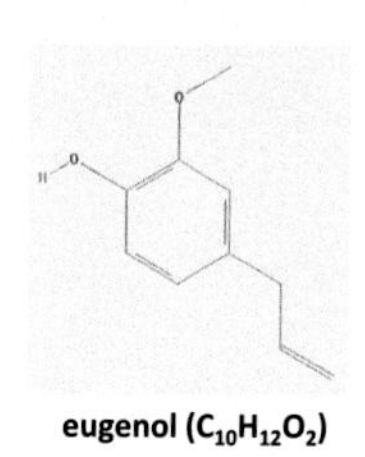

eugenol ($C_{10}H_{12}O_2$)

Liquorice

Emotions

amused

bored
worried

LIQUORICE
Definition

The aromas associated with sweet candies

Key aroma compound

anethole

Chocolate tasting

The sweetness of liquorice is very different from pure sugar, being less instant, tart, and lasting longer. This ensures that the chocolate will receive a different type of sweetness which will give the impression of the famous candy called 'drop'.

Grenada origin chocolate has a liquorice-like note coming from cocoa liquor.

Liquorice flavor

Liquorice has sweet, fruity notes associated with the root of 'Glycyrrhiza glabra'.

The aroma of liquorice comes from a complex, variable combination of compounds, of which anethole is up to 3% of total volatiles.

Liquorice smell

In almost every part of the world you can find this easily recognizable smell that is sweet, lingering, spicy and anise-like.

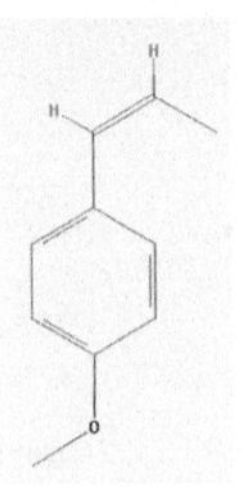

anethole ($C_{10}H_{12}O$)

DOWN-TO-EARTH

Olive

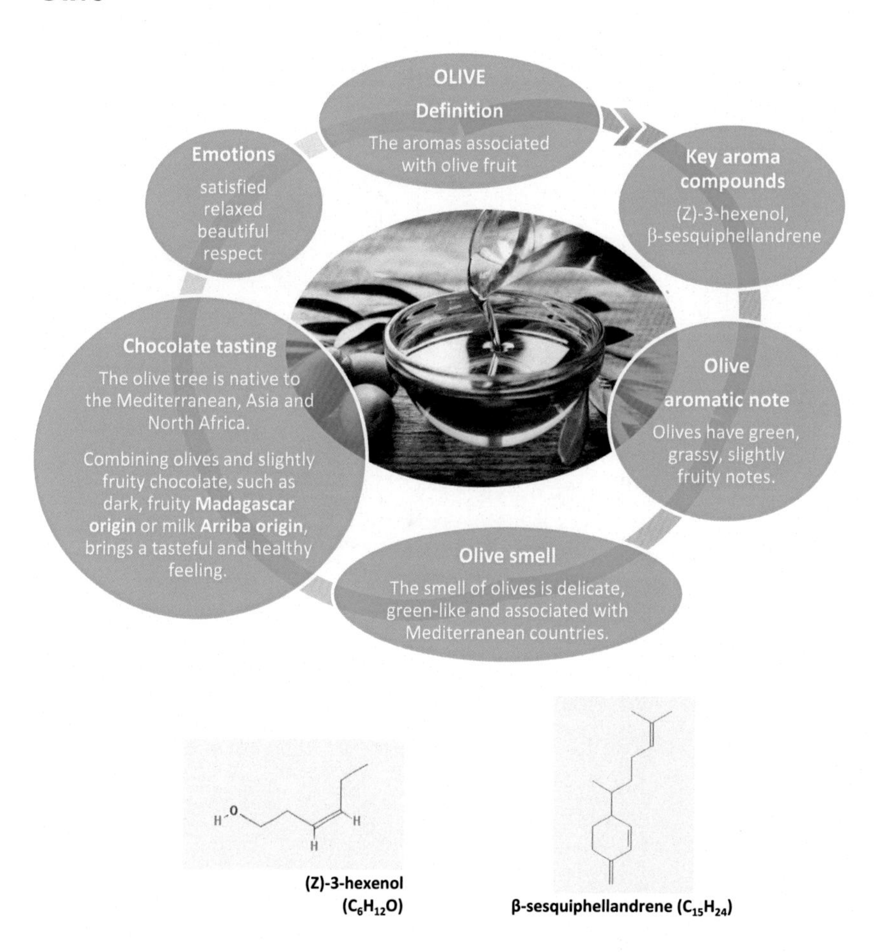

(Z)-3-hexenol ($C_6H_{12}O$)

β-sesquiphellandrene ($C_{15}H_{24}$)

Mushroom

MUSHROOM

Definition

The aromas associated with fresh mushrooms

Key aroma compound

cis-3-octen-1-ol

Mushroom aromatic note

Mushrooms have mealy, bleach-like notes and many interesting 'forest-like' nuances.

Mushroom smell

Depending on the mushroom variety, their smell may include: corn-like, farinaceous notes, swamp, coal-like or green almond nuances.

Chocolate tasting

Pairing mushrooms with chocolate is challenging and innovative.

Fruity, milk **Java** chocolate has very delicate notes of mushrooms.

Emotions

friendly
happy
amused

passive

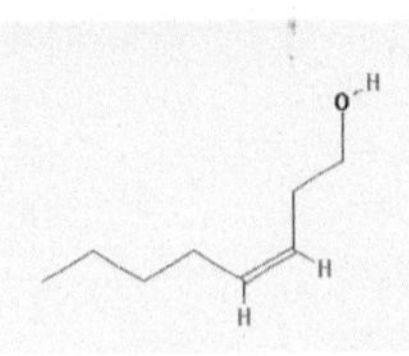

cis-3-octen-1-ol ($C_8H_{16}O$)

Tobacco

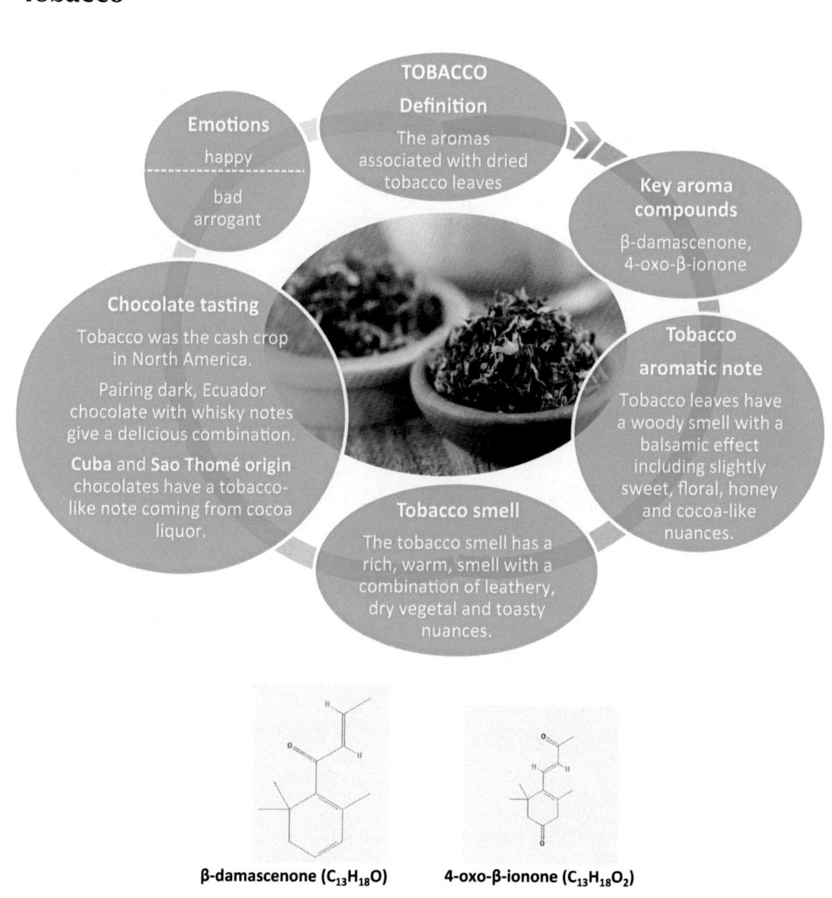

β-damascenone ($C_{13}H_{18}O$) **4-oxo-β-ionone ($C_{13}H_{18}O_2$)**

Rummy

RUMMY

Definition

The aromas associated with rum

Key aroma compounds

3-methyl-1-butanol, β-damascenone, 2-phenylethanol

Rummy flavor

Rum is generally quite sweet and contains all kind of shadows of caramel, spicy, clove, apple sauce, pepper and maple-like notes. It also has phenolic, balsamic and sometimes smoky flavor.

Rummy smell

The smell of rum has toasted, malty and also floral, rosy nuances.

Chocolate tasting

The origin of rum, a distilled alcoholic beverage made from sugarcane, is linked to Caribbean plantations in the 17th century.

Due to the aging process in oak barrels, it has a special woodiness that matches extremely well with fruity dark **Haiti** or **Ecuador origin** chocolate.

Emotions

friendly
amused
happy
powerful
strong

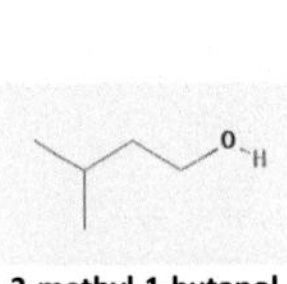

3-methyl-1-butanol (isoamyl alcohol) ($C_5H_{12}O$)

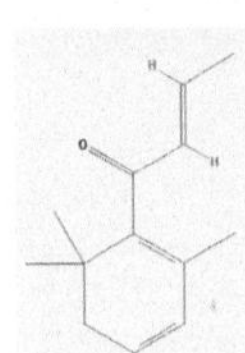

β-damascenone ($C_{13}H_{18}O$)

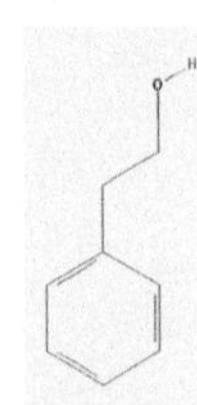

2-phenylethanol (phenethyl alcohol) ($C_8H_{10}O$)

Whisky

WHISKY

Definition

The aromas associated with whisky

Key aroma compound

whisky lactone

Whisky flavor

Whisky has green, malty, spicy, floral, woody, citrus, phenolic and peaty notes.

Whisky smell

A warm smell of whisky is intriguing and complex. Some people compare it to oriental scent, others to autumn-winter season.

Chocolate tasting

Green, malty, spicy aromas of whisky and roasted flavors of chocolate infuse in great harmony.

Dark, **Ecuador origin** chocolate pairs well with less salty whisky, while dark **Inaya** chocolate goes perfectly with 16 years old, salty Lagavulin.

Emotions

confident
friendly
happy

---

aggresive

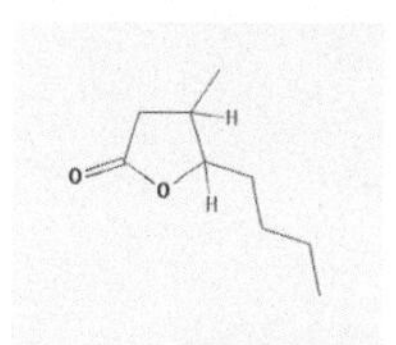

whisky lactone ($C_9H_{16}O_2$)

Woody oak

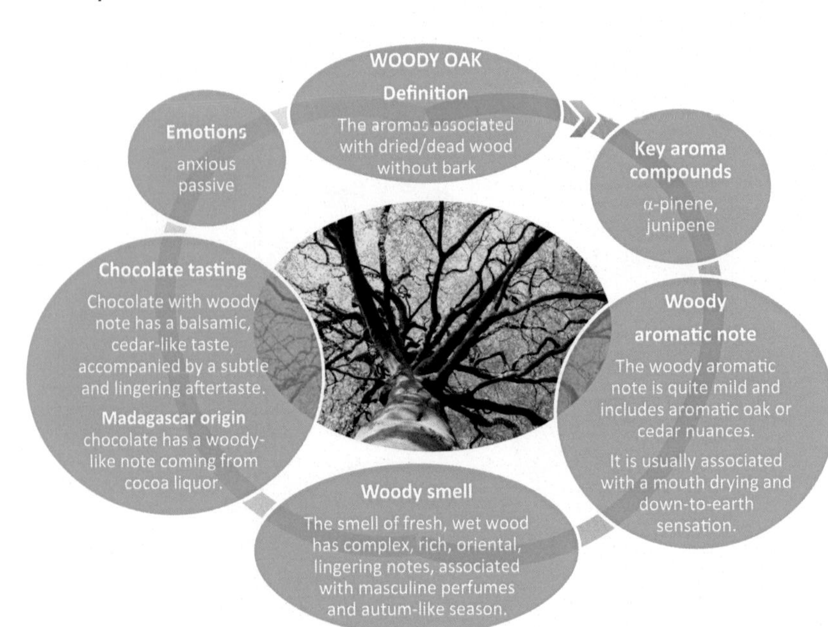

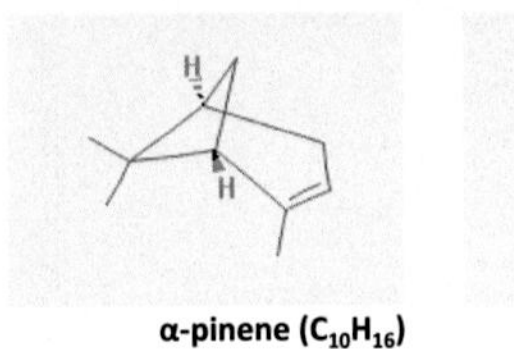

α-pinene ($C_{10}H_{16}$)

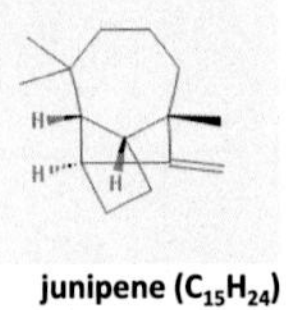

junipene ($C_{15}H_{24}$)

Smoked wood

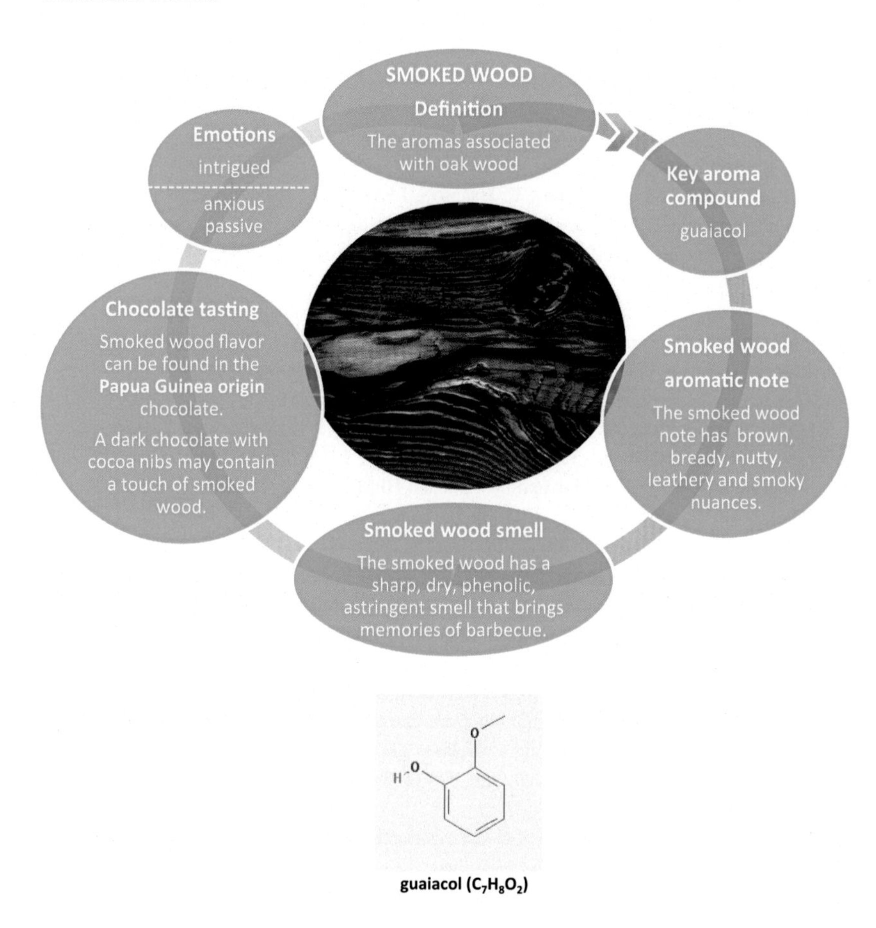

guaiacol ($C_7H_8O_2$)

FURTHER READING

Atanasova, B., Thomas-Danguin, T., Langlois, D., Nicklaus, S., Chabanet, C., Etiévant, P., 2005. Perception of wine fruity and woody notes: influence of peri-threshold odorants. Food Qual. Prefer. 16 (6), 504–510, (woody).

Bednarczyk, A.A., Kramer, A., 1975. Identification and evaluation of the flavor-significant components of ginger essential oil. Chem. Senses 1 (4), 377–386, (ginger).

Belitz, H.D., Grosch, W., Schieberle, P., 2009. Food Chemistry. Springer Science & Business Media, Germany, (vanilla, mushroom, oaky).

Brunschwig, C., Rochard, S., Pierrat, A., Rouger, A., Senger-Emonnot, P., Georgec, G., Raharivelomananaa, P., 2016. Volatile composition and sensory properties of Vanilla × tahitensis bring new insights for vanilla quality control. J. Sci. Food Agric. 96 (3), 848–858, (vanilla).

Burnside, E. (2012). Master thesis: characterisation of volatiles in commercial and self-prepared rum ethers and comparison with key aroma compounds of rum. University of Illinois at Urbana-Champaign, p. 57. (rummy).

Camara, J.S., Marques, J.C., Perestrelo, R.M., Rodrigues, F., Oliveira, L., Andrade, P., Caldeira, M., 2007. Comparative study of the whisky aroma profile based on headspace solid phase microextraction using different fibre coatings. J. Chromatogr. A 1150, 198–207, (whisky).

Castro-Vázquez, L., Díaz-Maroto, M.C., González-Viñas, M.A., Pérez-Coello, M.S., 2009. Differentiation of monofloral citrus, rosemary, eucalyptus, lavender, thyme and heather honeys based on volatile composition and sensory descriptive analysis. Food Chem. 112 (4), 1022–1030, (lavender, rosemary).

Eggink, P.M., Maliepaard, C., Tikunov, Y., Haanstra, J.P.W., Bovy, A.G., Visser, R.G.F., 2012. A taste of sweet pepper: volatile and non-volatile chemical composition of fresh sweet pepper (*Capsicum annuum*) in relation to sensory evaluation of taste. Food Chem. 132 (1), 301–310, (pepper).

Fujimaki, M., Kim, K., Kurata, T., 1974. Analysis and comparison of flavor constituents in aqueous smoke condensates from various woods. Agric. Biol. Chem. 38 (1), 45–52, (smoked wood).

Glabasna, A., Hofmann, T., 2007. Identification and sensory evaluation of dehydro- and deoxy-ellagitannins formed upon toasting of oak wood (*Quercus alba* L.). J. Agric. Food Chem. 55 (10), 4109–4118, (oaky).

Gutierrez, F., Albi, M.A., Palma, R., Rios, J.J., Olias, J.M., 1989. Bitter taste of virgin olive oil: correlation of sensory evaluation and instrumental HPLC analysis. J. Food Sci. 54 (1), 68–70, (olive).

http://www.bojensen.net/EssentialOilsEng/EssentialOils.htm. (rose, violet, clove, cinnamon, rosemary, coriander, tobacco, truffle).

Ito, Y., Kubota, K., 2005. Sensory evaluation of the synergism among odorants present in concentrations below their odor threshold in a Chinese jasmine green tea infusion. Mol. Nutr. Food Res. 49 (1), 61–68, (jasmine).

Jirovetz, L., Buchbauer, G., Ngassoum, M.B., Geissler, M., 2002. Aroma compound analysis of *Piper nigrum* and *Piper guineense* essential oils from Cameroon using solid-phase microextraction-gas chromatography, solid-phase microextraction-gas chromatography-mass spectrometry and olfactometry. Handbook of Essential Oils, 976 (1–2), Taylor & Francis Group, London, pp. 265–275, (pepper).

Kesen, S., Kelebek, H., Selli, S., 2014. Characterization of potent odorant compounds in Turkish olive oils. Int. J. Food Stud. 3, 248–257, (olive).

Koksal, N., Aslancan, H., Sadighazadi, S., Kafkas, E., 2015. Chemical investigation on Rose damascene Mill. Acta Sci. Pol. Hortorum Cultus 14 (1), 105–114, (rose).

Lawless, L.J.R., Hottenstein, A., Ellingsworth, J., 2012. The McCormick Spice Wheel: a systematic and visual approach to sensory lexicon development. J. Sens. Stud. 27 (1), 37–47, (clove).

Maga, J.A., 1981. Mushroom flavor. J. Agric. Food Chem. 29 (1), 1–4, (mushroom).

Miyazawa, M., Yoshinag, S., Kashima, Y., Nakahashi, H., Hara, N., Nakagawa, H., Usami, A., 2015. Chemical composition and characteristic odor compounds in essential oil from Alismatis Rhizoma (Tubers of Alisma orientale). J. Oleo Sci. 65 (1), 91–97, (woody).

Moore, J.G., Straight, R.C., Osborne, D.N., Wayne, A.W., 1985. Olfactory, gas chromatographic and mass-spectral analyses of fecal volatiles traced to ingested licorice and apple. Biochem. Biophys. Res. Commun. 131 (1), 339–346, (liquorice).

Muchtaridi, A., Subarnas, A., Apriyantono, A., 2010. Identification of compounds in the essential oil of nutmeg seeds (*Myristica fragrans* Houtt) that inhibit locomotor activity in mice. Essent. Oil Saf. 11, 4771–4781, (nutmeg).

Podlejski, J., Olejniczak, W., 1983. Methods and techniques in the research of tobacco flavor. Food/Nahrung 27 (5), 429–436, (tobacco).

Poisson, L., Schieberle, P., 2008. Characterization of the key aroma compounds in an American Bourbon Whisky by quantitative measurements, aroma recombination, and omission studies. J. Agric. Food Chem. 56 (14), 5820–5826, (whisky).

Prakash, M., Dattatreya, A., Bhat, K.K., 2003. Sensory flavor profiling and mapping of regional varieties of coriander (*Coriandrum sativum* L.). J. Sens. Stud. 18 (5), 409–422, (coriander).

Prashar, A., Locke, I.C., Evans, C.S., 2004. Cytotoxicity of lavender oil and its major components to human skin cells. Cell Prolif. 37 (3), 221–229, (lavender).

Ravi, R., Prakash, M., Bhat, K.K., 2007. Aroma characterisation of coriander (*Coriandrum sativum* L.). Eur. Food Res. Technol. 225, 367–374, (coriander).

Rétiveau, A.N., Chambers, I.V.E, Millken, G.A., 2004. Common and specific effects of fine fragrances on the mood of women. J. Sens. Stud. 19, 373–394, (woody).

Roueché, B., 1963. Alcohol in human culture. Lucia, Salvatore, P. (Eds.), Alcohol and Civilization, 178, McGraw-Hill, New York, (rummy).

Sharp, M.D. (2009). Analysis of vanilla compounds in vanilla extracts and model vanilla ice cream mixes using novel technology. Master Thesis. The Ohio State University, p. 162. (vanilla, vanillin).

Takahashi, M., Inai, Y., Miyazawa, N., Kurobayashi, Y., Fujita, A., 2013. Key odorants in cured Madagascar vanilla beans (*Vanilla planiforia*) of differing bean quality. Biosci. Biotechnol. Biochem. 77 (3), 606–611, (vanillin).

Vahdatzadeh, M., Deveau, A., Splivallo, R., October 2015. The role of the microbiome of truffles in aroma formation: a meta-analysis approach. Appl. Environ. Microbiol. 81 (20), 6946–6952, (truffles).

Van Ruth, S.M., Roozen, J.P., 1994. Gas chromatography/sniffing port analysis and sensory evaluation of commercially dried bell peppers (*Capsicum annuum*) after rehydration. Food Chem. 51 (2), 165–170, (pepper).

Zhen, Y.-S., 2002. Tea: Bioactivity and Therapeutic Potential. Taylor & Francis, London, (jasmine).

Chapter 7

Trigeminal Effects

Chapter Outline

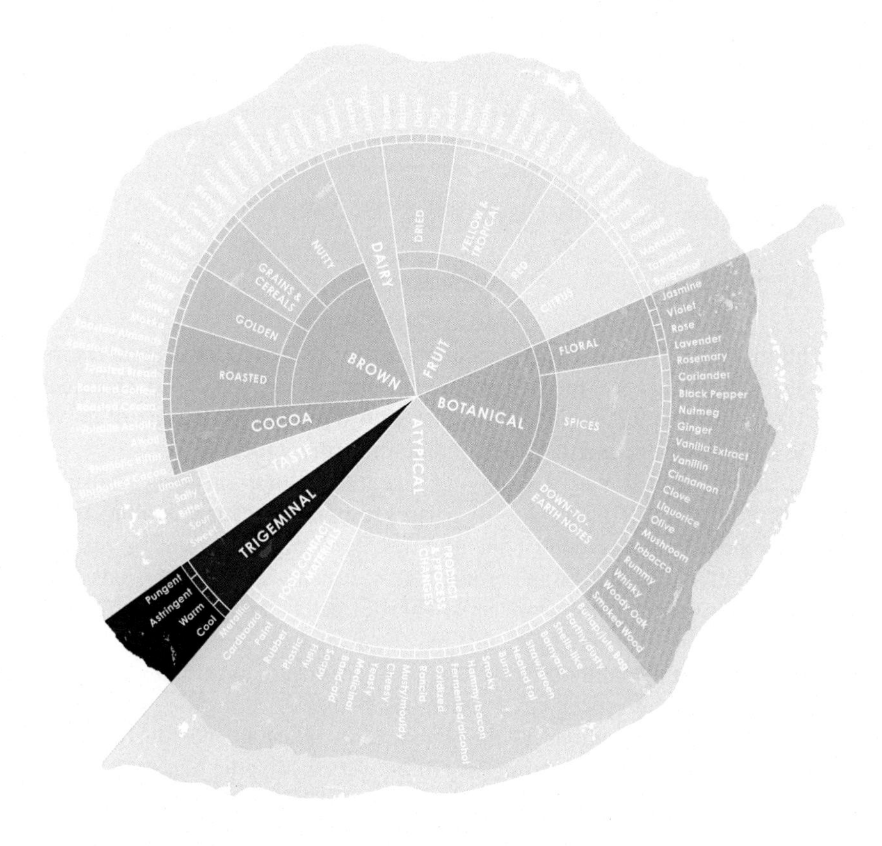

Hidden Persuaders in Cocoa and Chocolate. http://dx.doi.org/10.1016/B978-0-12-815447-2.00008-5

Science shows that a common metaphor of something tasting or feeling as 'cool' or 'warm' is not in the cultural background of the taster or a specific part of the brain sensing. Actually, it is encoded within the sensor molecules in the nerve endings of the skin and one of the most important nerves located in the head called the trigeminal nerve.

The trigeminal nerve is the fifth cranial nerve or simply CN V and is responsible for the sensation in the face and motor functions such as biting and chewing. It is called trigeminal because it splits to three major branches: the ophthalmic nerve (V1), the maxillary nerve (V2) and the mandibular nerve (V3). The ophthalmic and maxillary nerves are purely sensory, and the mandibular nerve has sensory (or "cutaneous") and motor functions.

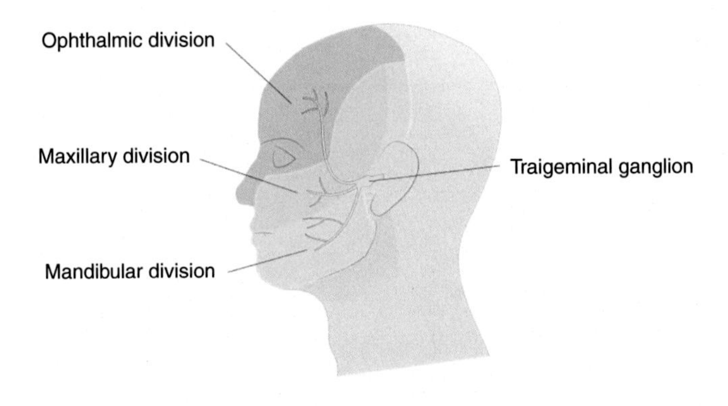

Trigeminal nerve

The trigeminal nerve cells are called 'sensory receptors'. These receptors include thermoreceptors, mechanoreceptors, chemoreceptors (send chemical signals) and nociceptors (pain related) that send signals along a chain of nerve cells to the spinal cord and then to the brain for further processing.

It was established that a family of the temperature-sensitive Transient Receptor Potential (TRP) receptors or sensors respond to heating, cooling, pungent and other sensations found in plants. These sensors are triggered by various chemical compounds activating certain TRP's, causing an influx of ions into the heat-sensitive interior of nerve cells. This normally starts a cascade of reactions leading to a sensation of pain that may be actually quite pleasant for some people because of the cultural factor, that is, the exposure to certain spices and ingredients in a certain social environment.

The trigeminal sensations in this publication relate to the four selected effects of *cooling, warming, astringency or pungency*. Metallic sensation is also classified as trigeminal but since it is considered 'atypical' in cocoa and chocolate product, therefore it is discussed in the following chapter.

Thermo sensors, active during cooling perception, are named TRPM8. *Cooling* effect is often associated with products flavored with menthol found in mint or eucalyptol from eucalyptus. Research shows that different forms of menthol isomers bring different intensity and duration of cooling effect. Compared to d-menthol, I-menthol isomer samples exhibit a greater maximum intensity and longer total duration of cooling. Hereby, the burning sensation increases with concentration. Compounds that bring a cooling effect are often used in skin creams, mouthwash, refreshing drops or throat soothing lozenges.

Thermo sensors responsible for warmth perception are called TRPV1. The *warming* effect comes from substances, such as capsaicin found in paprika, or camphor present in nutmeg or cinnamon. It is interesting to know that birds cannot detect capsaicin because they swallow the whole seeds that are mixed by people with chili peppers, in order to deter other animals from stealing them. It is also known that some animals have high heat-sensitivity or even infrared radiation sensitivity, like vampire bats, while other mammals do not have this ability anymore. It is explained that humans have gradually lost this trigeminal over-sensitivity due to the evolution process.

Astringency is linked to the presence of tannic acid and other substances. It was extensively studied and linked to a complex perception including: mouth drying, mouth roughing, pucker feeling, as well as some bitterness and sourness. It was also found that tasting of the astringent compounds results in the formation of soluble protein–phenol complexes.

Pungency is linked to the presence of a chemical called allyl isothiocyanate (AITC) found in wasabi, horseradish or yellow mustard. This compound activates a different sensor of the TRP family called TRPA1.13. Other compounds in that group include allicin or DADS (Diallyl disulfide), that are found in garlic and onion. Pungency, rare in pure cocoa or chocolate products, may be however experienced from the ingredients paired with chocolate. Pungent qualities of red pepper, black pepper, cinnamon, cumin, cloves and ginger were investigated and it was established that they may be described in different words such as burning, tingling and numbing; and can have some skin and eye irritation. Researchers also found that pungent impressions have different time-lasting effect.

The intensity and the duration of these sensations have paramount importance in overall flavor perception. These aspects, like all flavors, can be perfectly measured by the sensory panel, using the Temporal Dominance of Sensations (TDS) method.

COOL

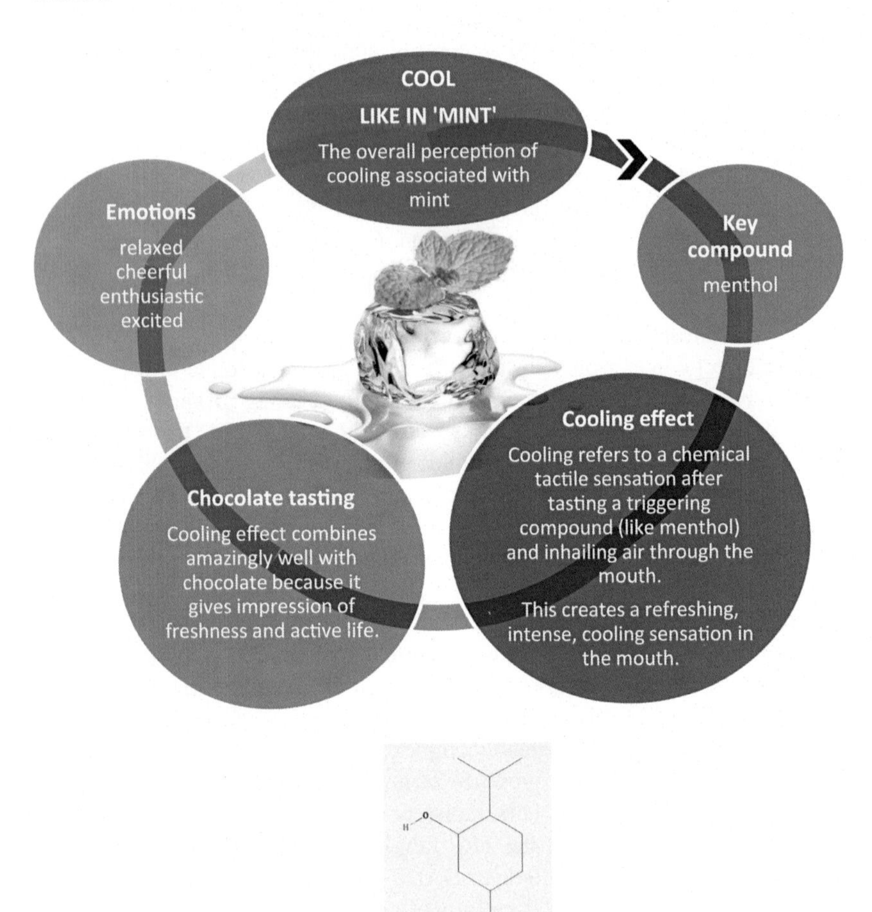

menthol ($C_{10}H_{20}O$)

WARM

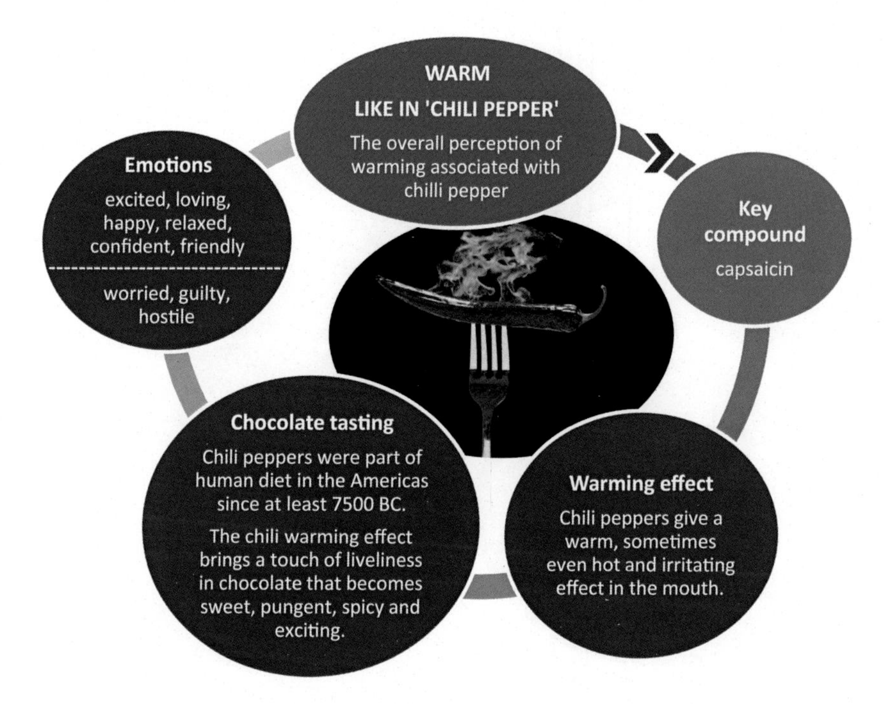

capsaicin ($C_{18}H_{27}NO_3$)

ASTRINGENT

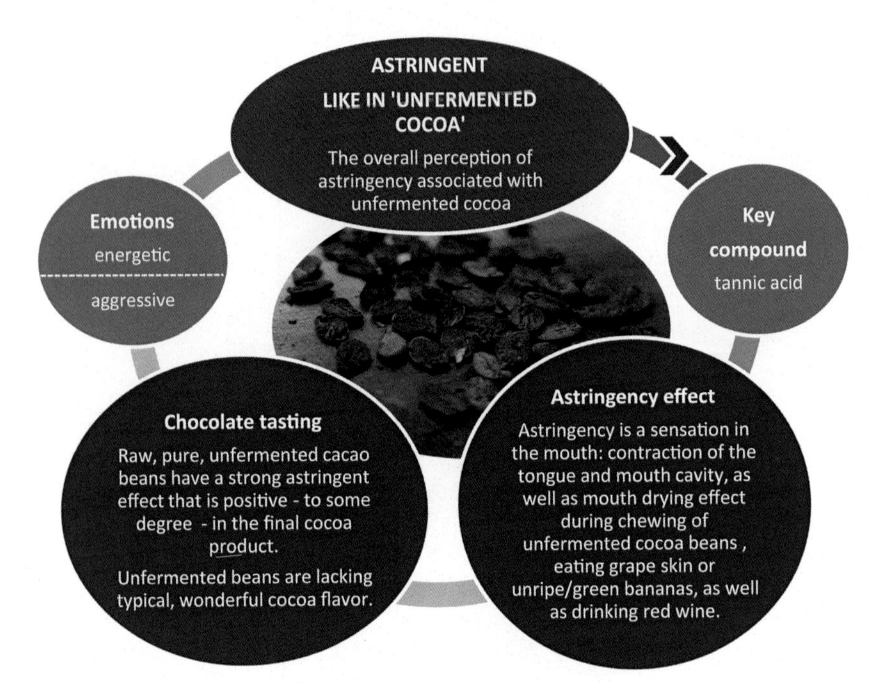

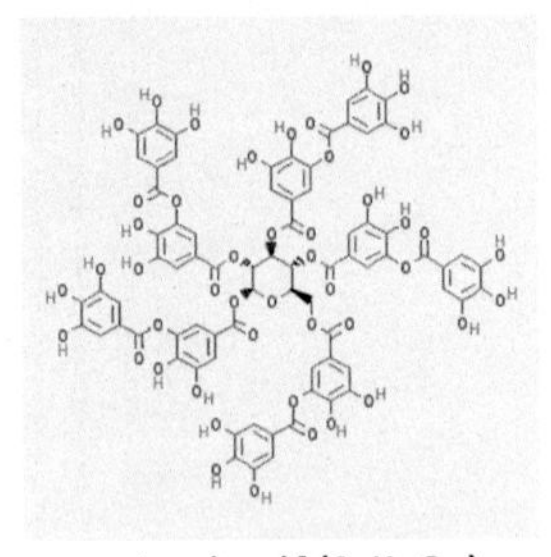

tannic acid ($C_{76}H_{52}O_{46}$)

PUNGENT

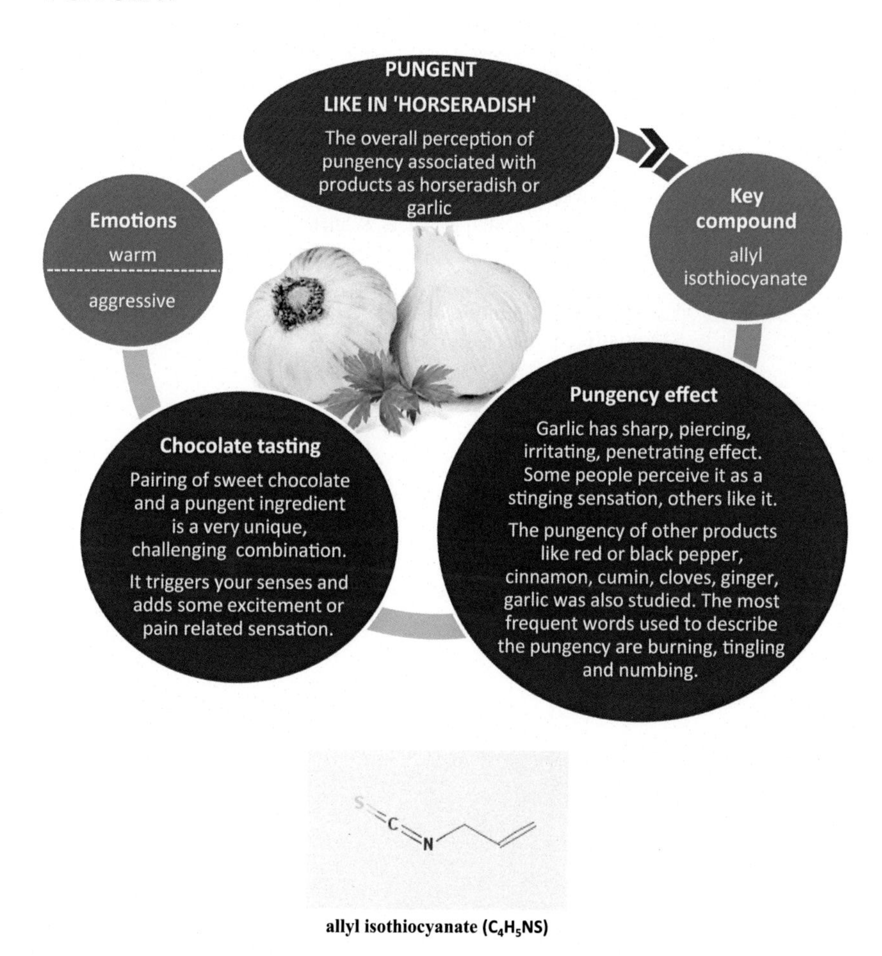

allyl isothiocyanate (C_4H_5NS)

FURTHER READING

Cliff, M., Heymann, H., 1992. Descriptive analysis of oral pungency. J. Sens. Stud. 7 (4), 279–290, (pungency).

Green, B.G., Alvarez-reeves, M., George, P., Akirav, C., 2005. Chemesthesis and taste: evidence of independent processing of sensation intensity. Physiol. Behav. 86, 526–537, (warming).

Gwartney, E., Heyman, H., 1995. The temporal perception of menthol. J. Sens. Stud. 10 (4), 393–400, (cooling).

Hamilton, J. (2015). Sushi Science: A 3-D View of the Body's Wasabi Receptor. NPR Health Shots, April 8. http://www.npr.org/sections/healthshots/2015/04/08/398065961/sushi-science-a-3-d-view-of-thebodyswasabi-receptor. (pungency).

Hayes, J.E., Feeney, E.L., Allen, A.L., 2013. Do polymorphisms in chemosensory genes matter for human ingestive behavior? Food Qual. Prefer. 30, 202–216, (astringency).

Kallithraka, S., Bakker, J., Clifford, M.N., 1998. Evidence that salivary proteins are involved in astringency. J. Sens. Stud. 13 (1), 29–43, (astringency).

Lee, C.B., Lawless, H.T., 1991. Time-course of astringent sensations. Chem. Senses 16 (3), 225–238, (astringency).

Linden, D.J. (2015). How We Sense the Heat of Chili Peppers and the Cool of Menthol. Incredible thermosensors that let you know the difference between the spicy hot of chili and the chilly sensation of mint. Scientific American, February, pp. 1–11. Excerpted from: Touch: The Science of Hand, Heart and Mind, by David J. Linden. Viking. (warming).

Pineau, N., Goupil de Bouillé, A., Lepage, M., Lenfant, F., Schlich, P., Martin, N., Rytz, A., 2012. Temporal dominance of sensations: what is a good attribute list? Food Qual. Prefer. 26, 159–165, (TDS).

Tominaga, Makoto, 2005. Molecular mechanisms of trigeminal nociception and sensation of pungency. Chem. Senses 30, 191–192, (pungency).

Chapter 8

Atypical

Chapter Outline

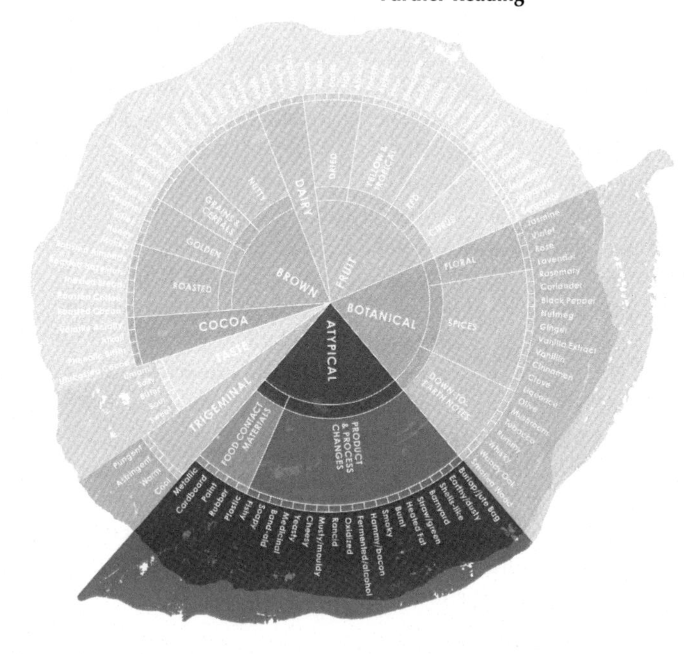

Hidden Persuaders in Cocoa and Chocolate. http://dx.doi.org/10.1016/B978-0-12-815447-2.00009-7

Atypical flavors in cocoa and chocolate are certainly not wanted from the quality point of view. Behind their presence are chemical compounds that stimulate intuitively feelings of dislike, disgust, disapproval, disinterest, sadness, tiredness; and sometimes create even more strong emotions, such as aggressiveness, fear or being scared.

Food chemists can perfectly analyze the presence of atypical flavors or off-flavors in cocoa and chocolate. Also a trained panel will have no problem to detect the off-flavor associated with fat oxidation that can be linked for instance to a high concentration of hexanal, or to indicate rancidity being activated by butanoic acid. Other examples include too high concentration of guaiacol, linked to smokiness, or ethyl acetate that may indicate a yeasty off-flavor. It is remarkable that all of these off-flavors in cocoa and chocolate products bring us an immediate impression of possible contamination.

Managing the quality of precious cocoa beans, and all products made from them, is truly a challenging task. All steps of producing, handling, processing and storage need to be considered with the same attention and carefulness. Quality control starts as a professional quality check at the farmer level, continues at every level of the supply chain, and finishes at the well-designed quality monitoring system located at the point of sale.

In general, most of flavor contamination in cocoa and chocolate can be classified into *two categories*: the first is related to natural product changes and process related changes, and the second is linked to the migration or permeation of volatile compounds from the packaging or the environment into the food product.

In the first category, *product/process related changes*, we find the following off-flavors: *burlap/jute bag, earthy/dusty, shells-like/cocoa bran, barnyard, straw/green, heated fat, burnt, smoky, hammy/bacon, fermented/alcohol, oxidized, rancid, musty/mouldy, cheesy, yeasty, medicinal, soapy, fishy.*

Product and process changes in cocoa or chocolate may occur at all levels of cocoa beans handling, bean processing, storage environment, transportation, chocolate processing, further storage, displaying at points of sale and even during the storage at the consumer's kitchen. These changes will be initiated by a wrong handling at each specific step. For instance, at the farmer level, a wrong fermentation process may initiate a so called hammy off-flavor, which cannot be removed from beans and will persist until the end of the cocoa liquor process. Other examples are, for example, a bad drying of cocoa beans leading to a smoky note, or having cocoa beans for a too long time into jute bags during transport, leading to a burlap off-flavor.

In the second category, there are off-flavors linked to contamination due to the *food contact materials (FCM)*. These include the following off-flavors: *plastic, rubber, paint, cardboard, metallic.* This short list can be extended depending on the used packaging material. Please, note that paint and cardboard off-flavors are mainly caused by fat oxidation volatiles.

All cocoa and chocolate final products require packaging. As products need to be sometimes repacked, the type of packaging and the re-packaging process are extremely important factors in maintaining high quality until the end consumer.

Transportation and storage is considered as another key factor in keeping a good quality product. Shipping beans from the equatorial region to the northern hemisphere takes several weeks. Preserving them in the harbors on both sides requires specialized tools and controlled storage conditions. Not only beans, but all other ingredients for chocolate manufacture shall be stored and regularly checked for quality. The cocoa and chocolate industry developed the program of combined analytical and sensorial quality checks applied to all ingredients. It is highly advisable to implement the sensory quality assurance program (SQAP) by a qualified panel at all production sites. Here follows a high-level description of the procedure: minimum three people give a score for each ingredient (with conformity tables for each available) and the calculated average of the 'quality score' is an indication whether the product may be allowed into the production or shall be rejected.

Additionally, it is interesting to notice that food chemists specializing in off-flavors and taints detection distinguish two kinds of routes of contamination. 'Systematic' off-flavors are linked to product's aging and oxidation, and 'accidental' off-flavors may come from the packaging, environment during transporting, improper processing – like drying or over-fermentation, or even batch-to-batch variability of raw materials.

Finally, we shall remember that there is nothing such as a perfect storage condition or perfect packaging. Time and humidity naturally fluctuate and these fluctuations may influence the content of the packaged goods. When conditions become extreme, like in tropic, wet, or very cold countries, a faster deterioration may occur. Such changes are reflected in accelerated oxidation of most sensitive products like cocoa butter, white or milk chocolate. One should sensitize those responsible for the selling points about the fact that exposure to light and temperature increases the chance of oxidation, in all cocoa and chocolate products. Further challenges relate to migration of negative chemical compounds from the primary or secondary packaging, and even from the surrounding environment; examples include woody pallets or cleaning materials stored in the same place.

PRODUCT AND/OR PROCESS CHANGES

Burlap/jute bag

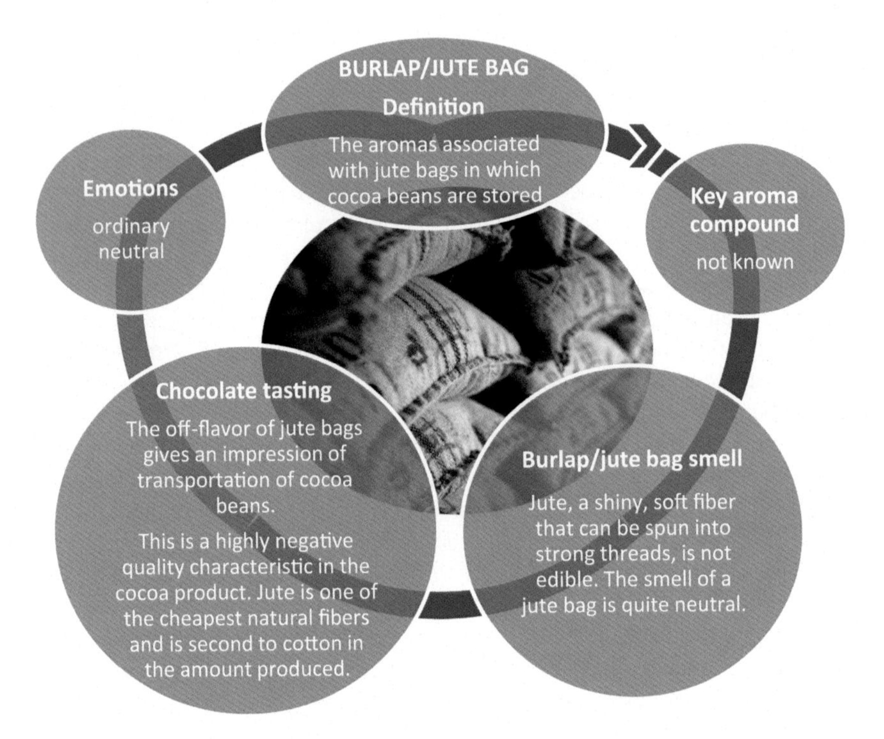

Earthy/dusty

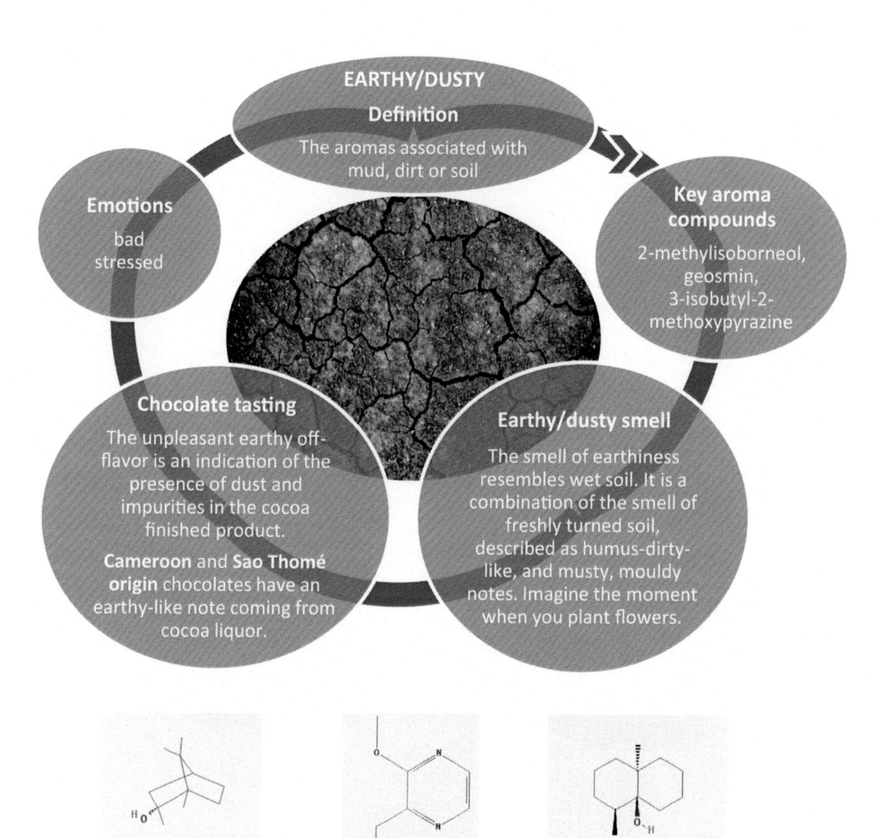

2-methylisoborneol ($C_{11}H_{20}O$)

3-isobutyl-2-methoxypyrazine ($C_9H_{14}N_2O$)

geosmin ($C_{12}H_{22}O$)

Shells-like

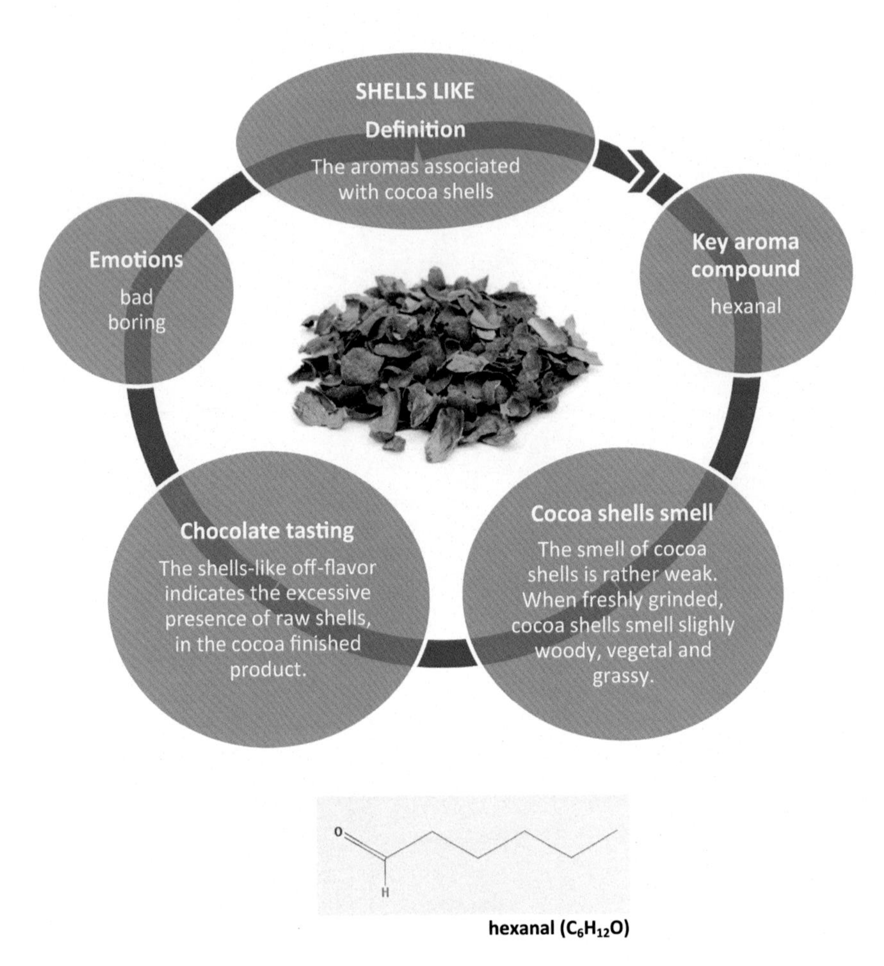

hexanal ($C_6H_{12}O$)

Barnyard

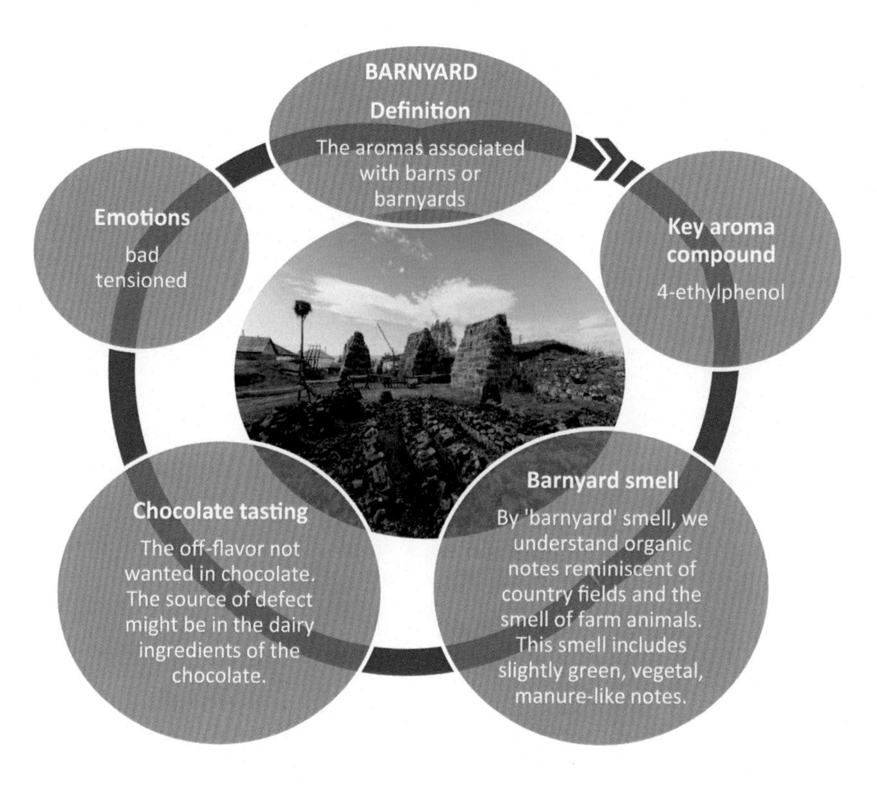

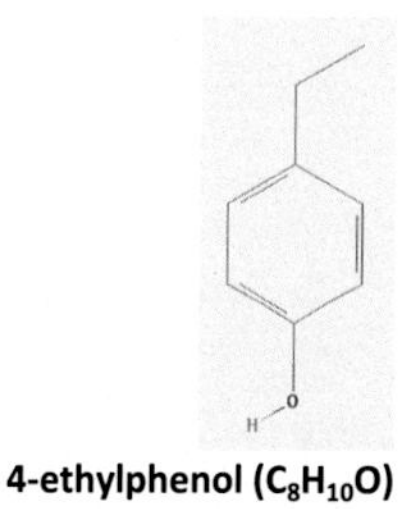

4-ethylphenol ($C_8H_{10}O$)

Straw/green

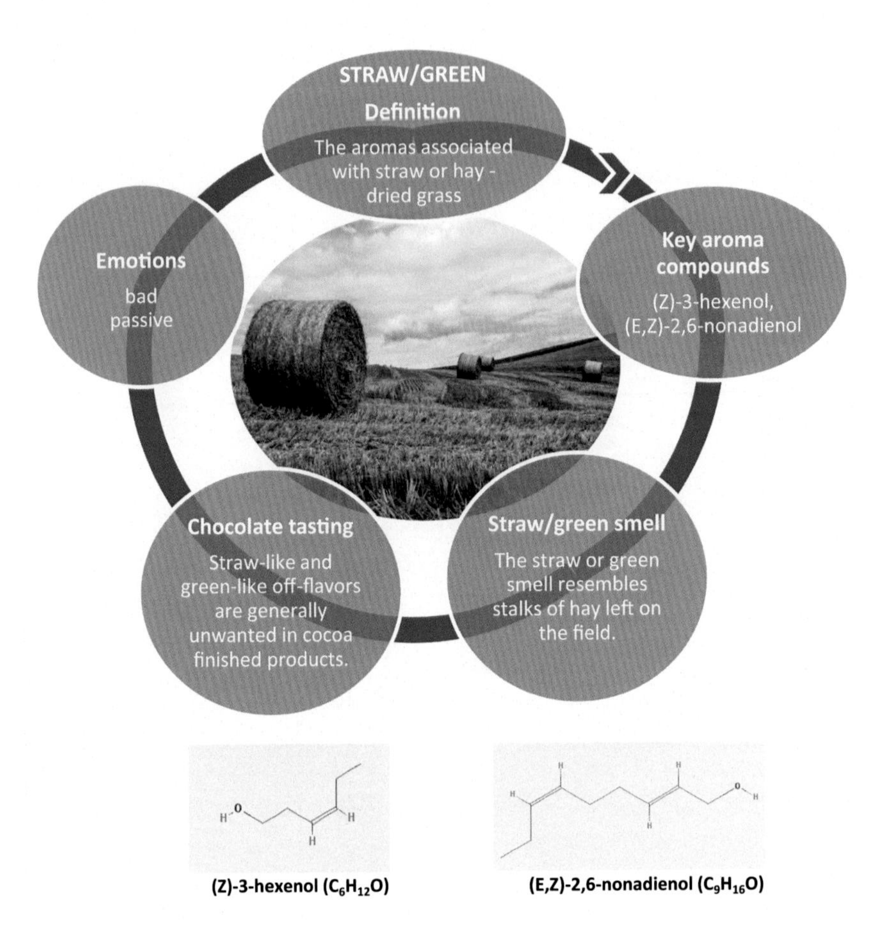

(Z)-3-hexenol ($C_6H_{12}O$)

(E,Z)-2,6-nonadienol ($C_9H_{16}O$)

Heated fat

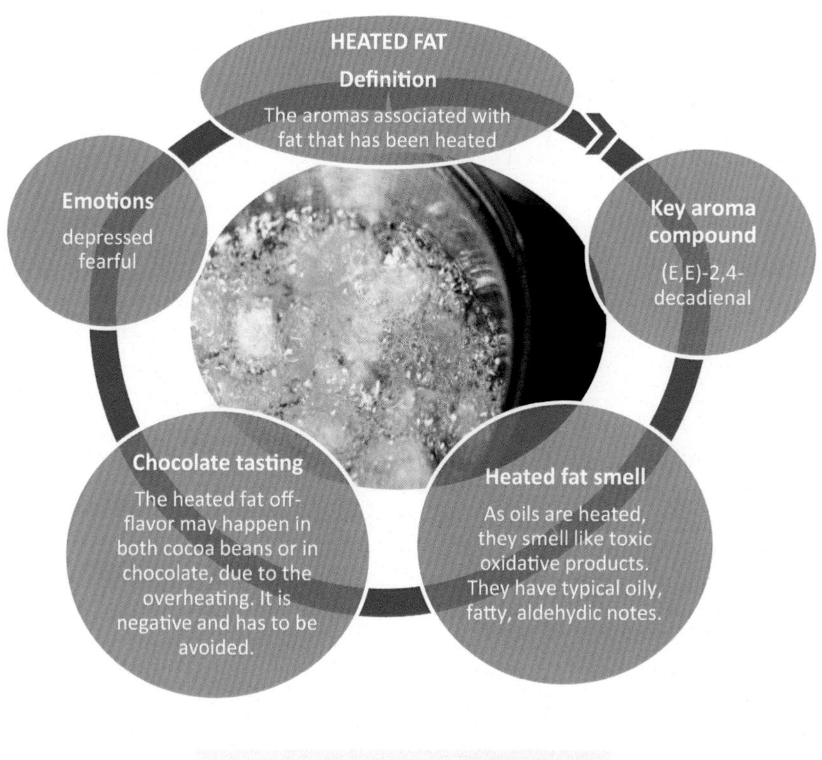

(E,E)-2,4-decadienal ($C_{10}H_{16}O$)

Burnt

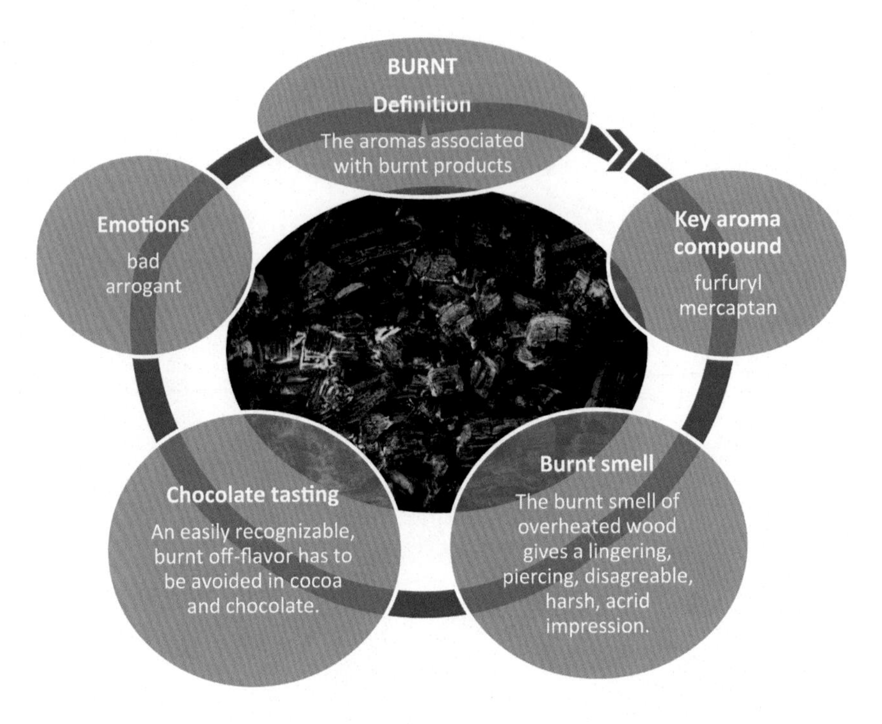

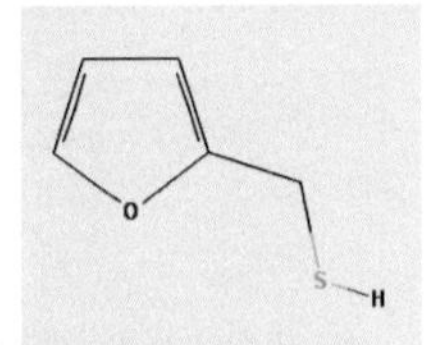

furfuryl mercaptan (C_5H_6OS)

Smoky

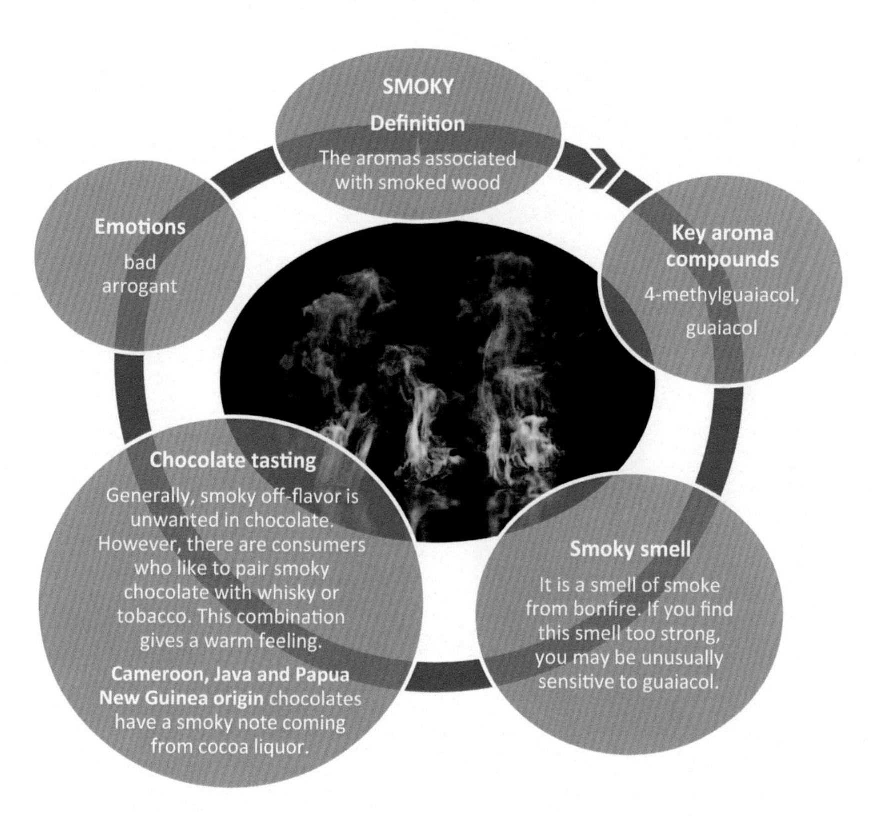

4-methylguaiacol ($C_8H_{10}O_2$)

guaiacol ($C_7H_8O_2$)

Hammy/bacon

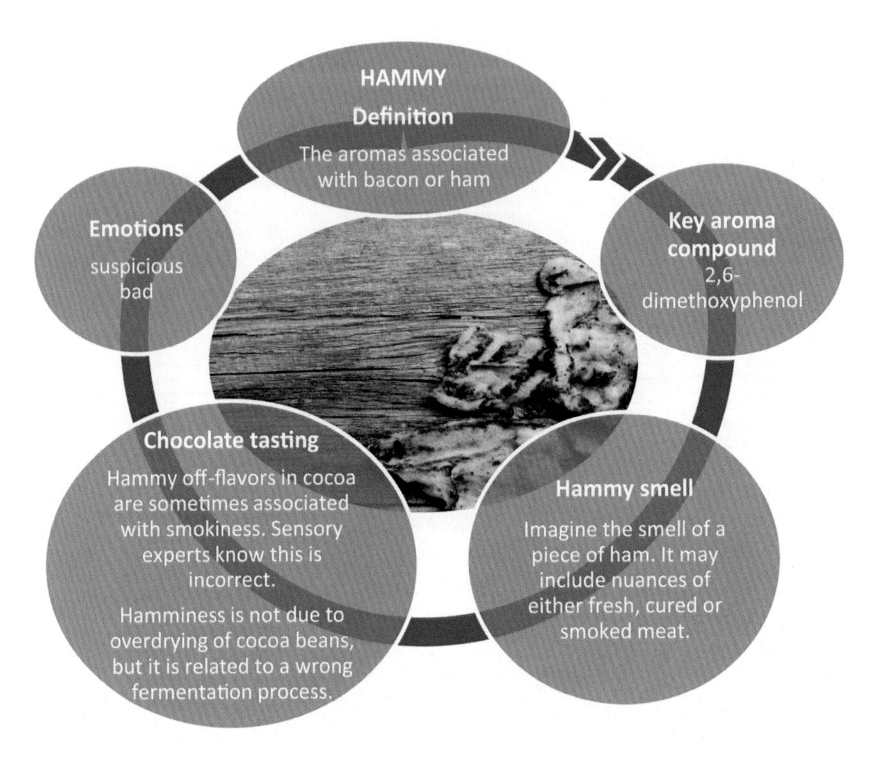

2,6-dimethoxyphenol ($C_8H_{10}O_3$)

Fermented/alcohol

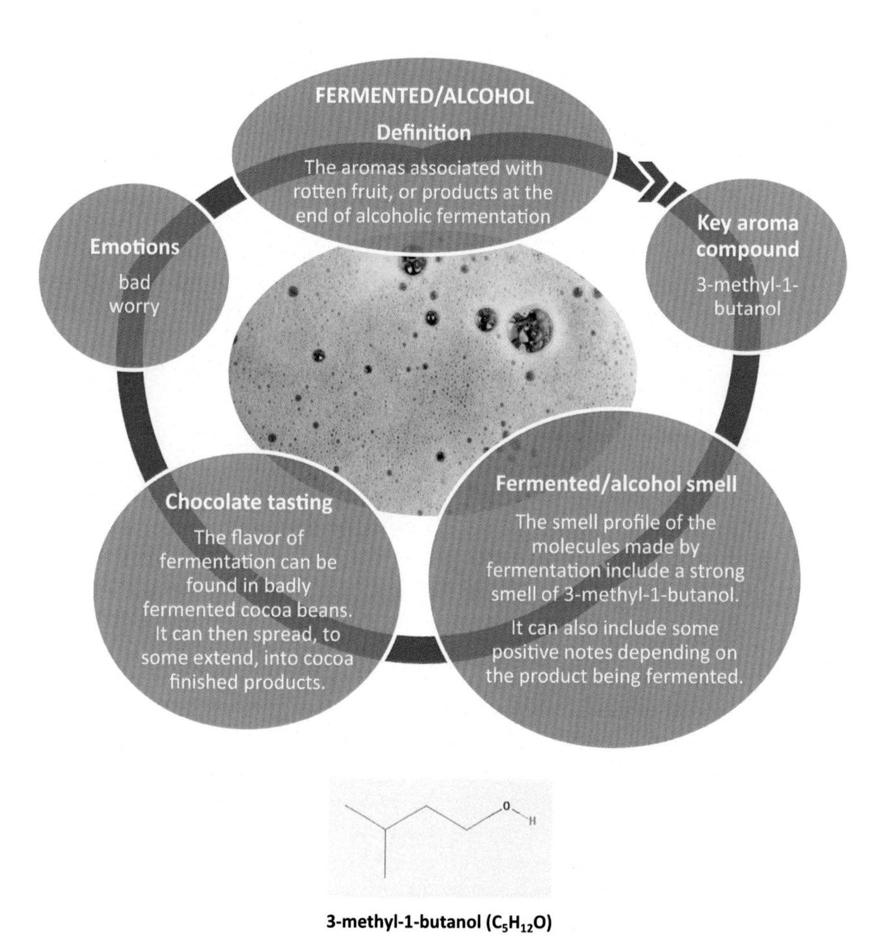

3-methyl-1-butanol ($C_5H_{12}O$)

Oxidized

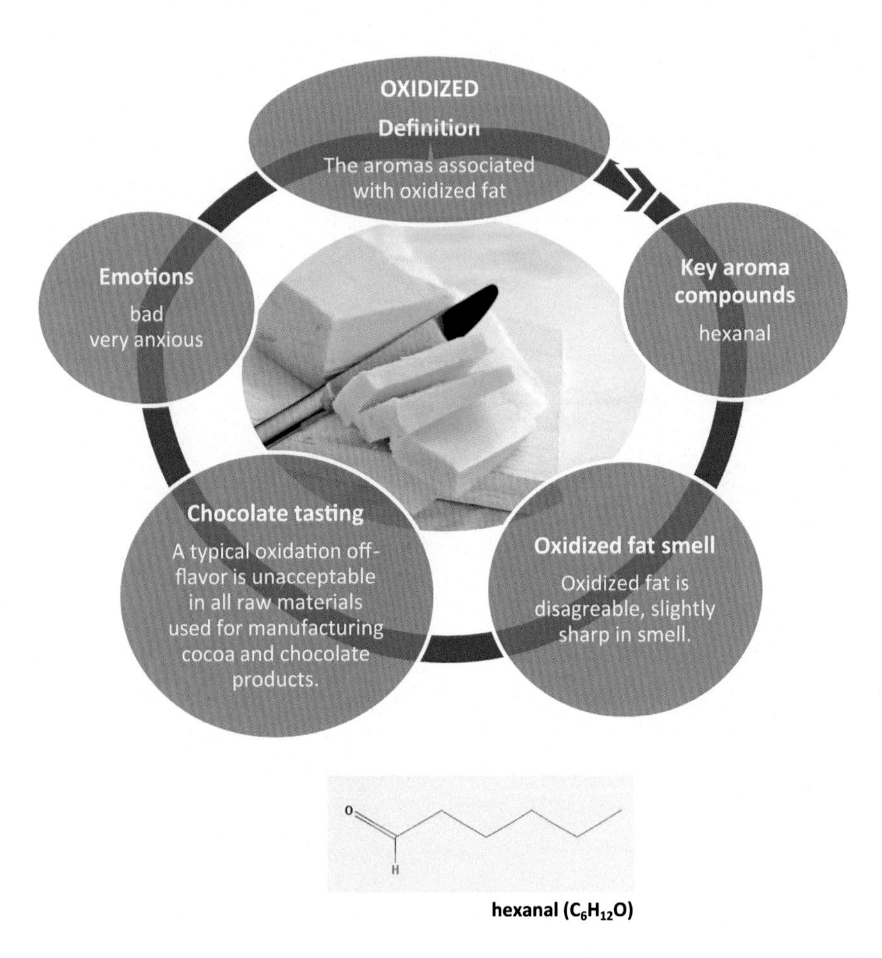

hexanal ($C_6H_{12}O$)

Rancid

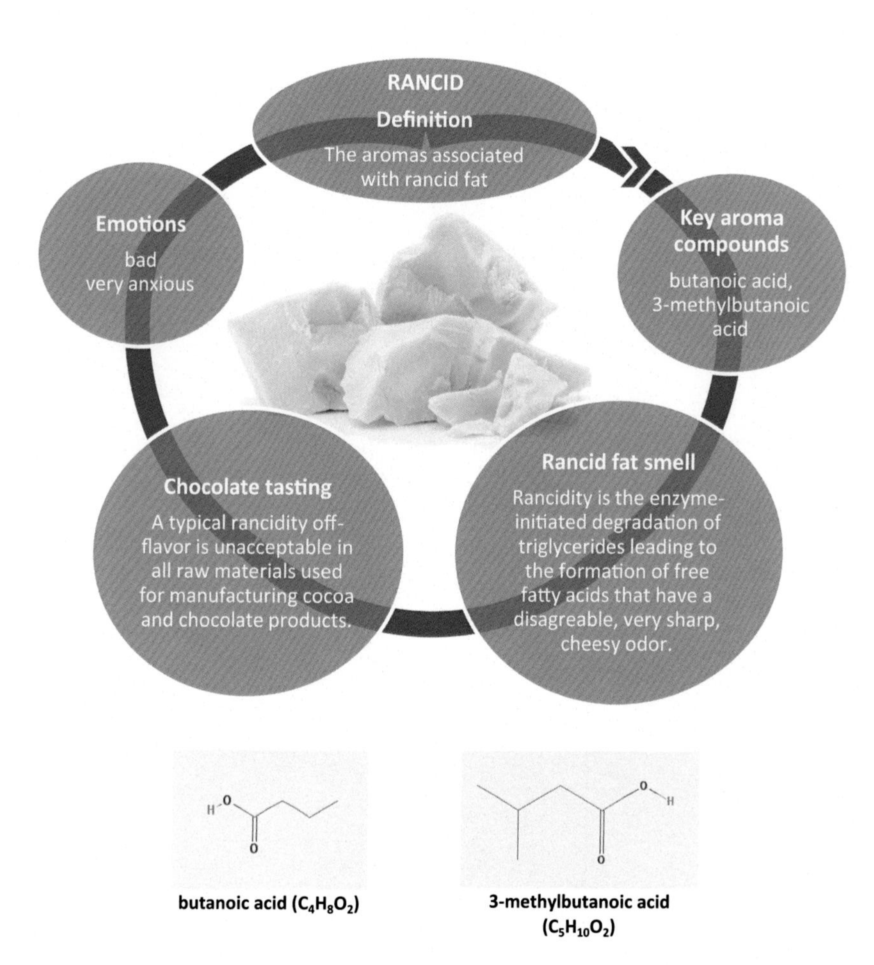

butanoic acid ($C_4H_8O_2$)

3-methylbutanoic acid ($C_5H_{10}O_2$)

Musty/mouldy

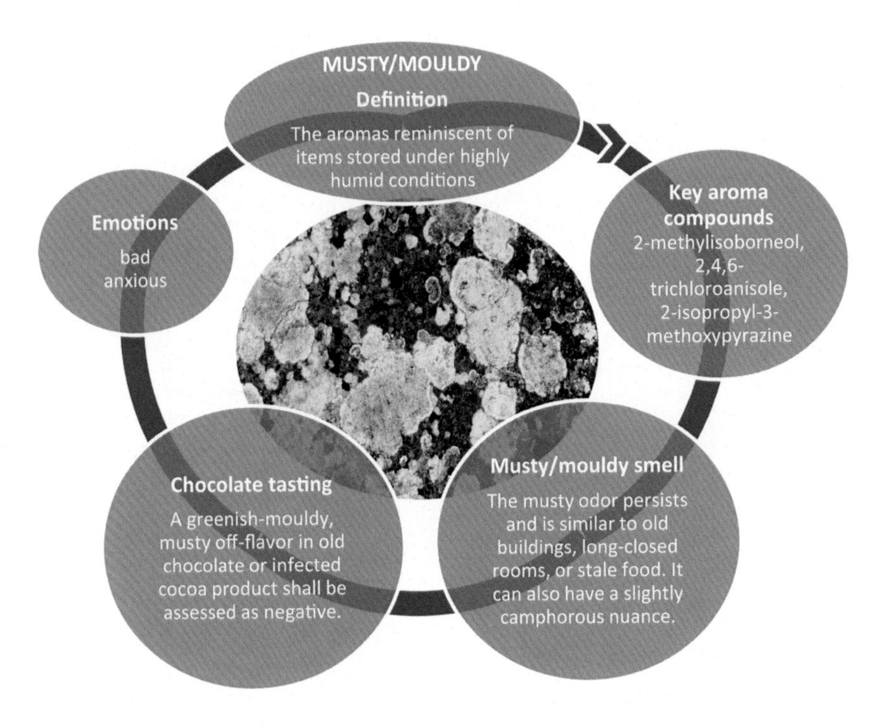

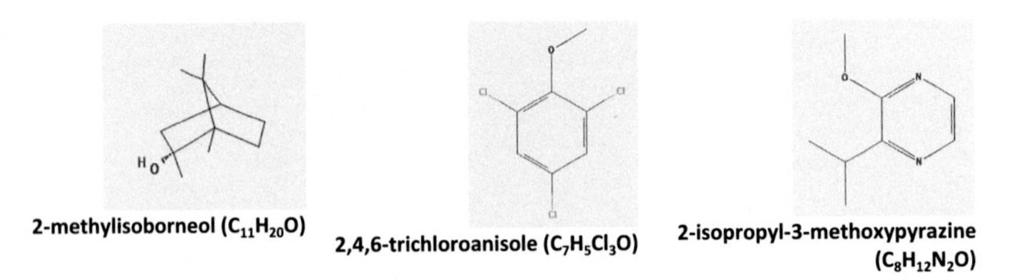

2-methylisoborneol ($C_{11}H_{20}O$) **2,4,6-trichloroanisole ($C_7H_5Cl_3O$)** **2-isopropyl-3-methoxypyrazine ($C_8H_{12}N_2O$)**

Cheesy

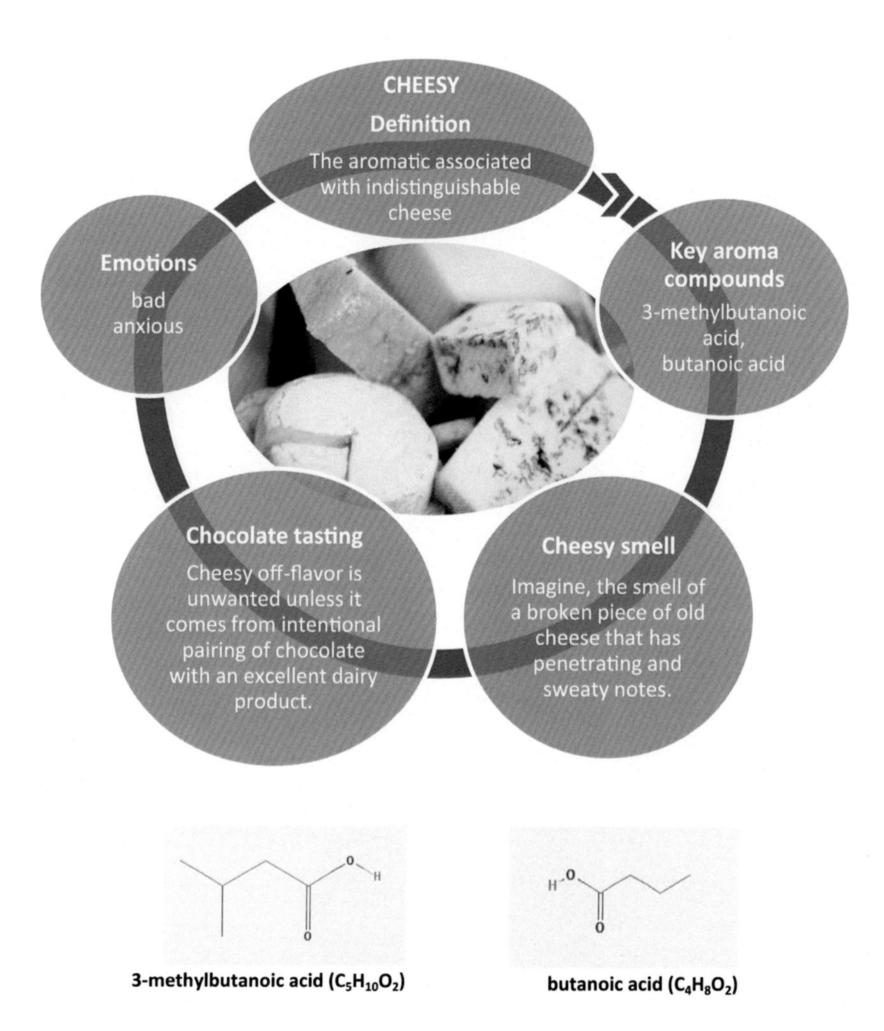

3-methylbutanoic acid ($C_5H_{10}O_2$)

butanoic acid ($C_4H_8O_2$)

Yeasty

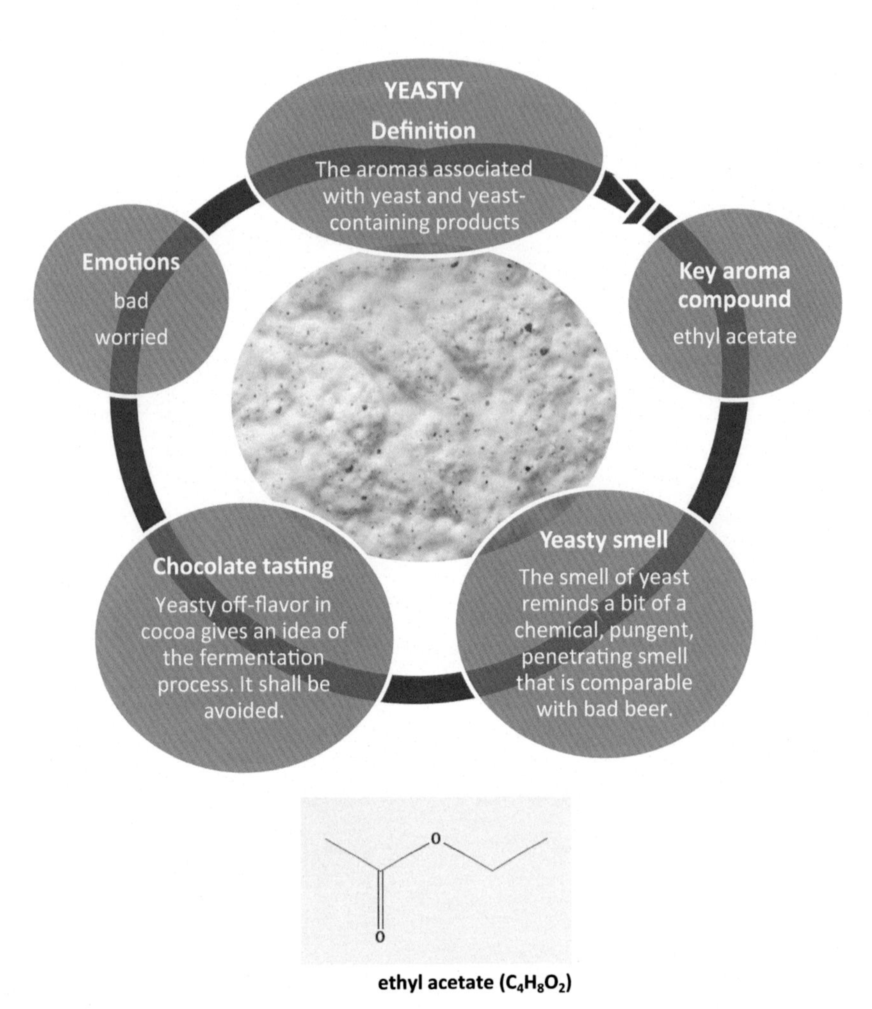

ethyl acetate ($C_4H_8O_2$)

Medicinal

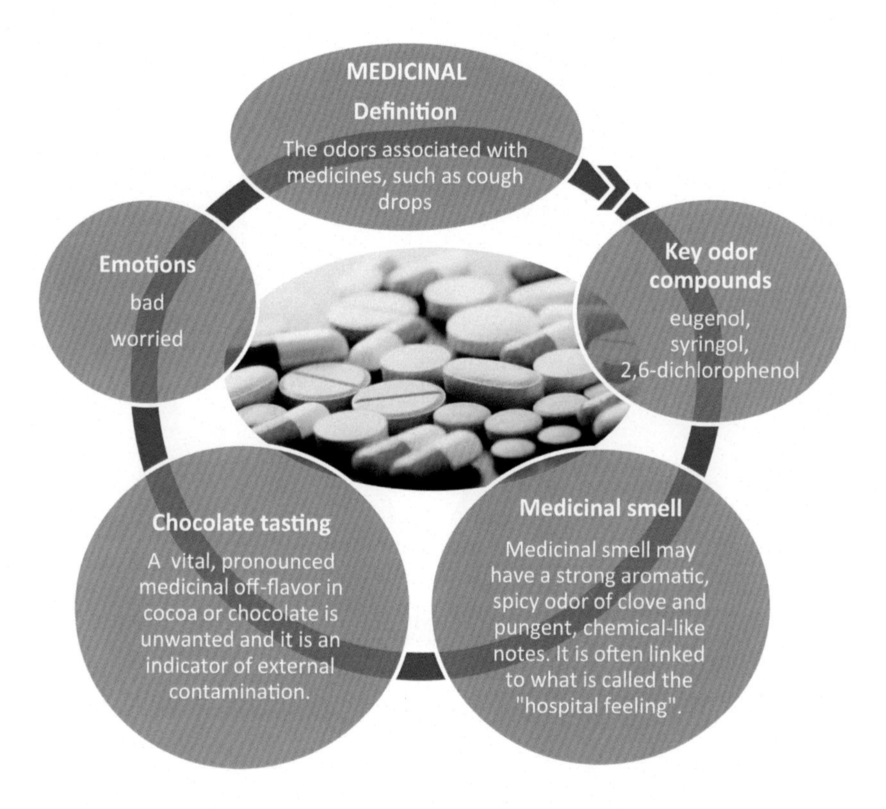

eugenol ($C_{10}H_{12}O_2$)

syringol ($C_8H_{10}O_3$)

2,6-dichlorophenol ($C_6H_4Cl_2O$)

Band-aid

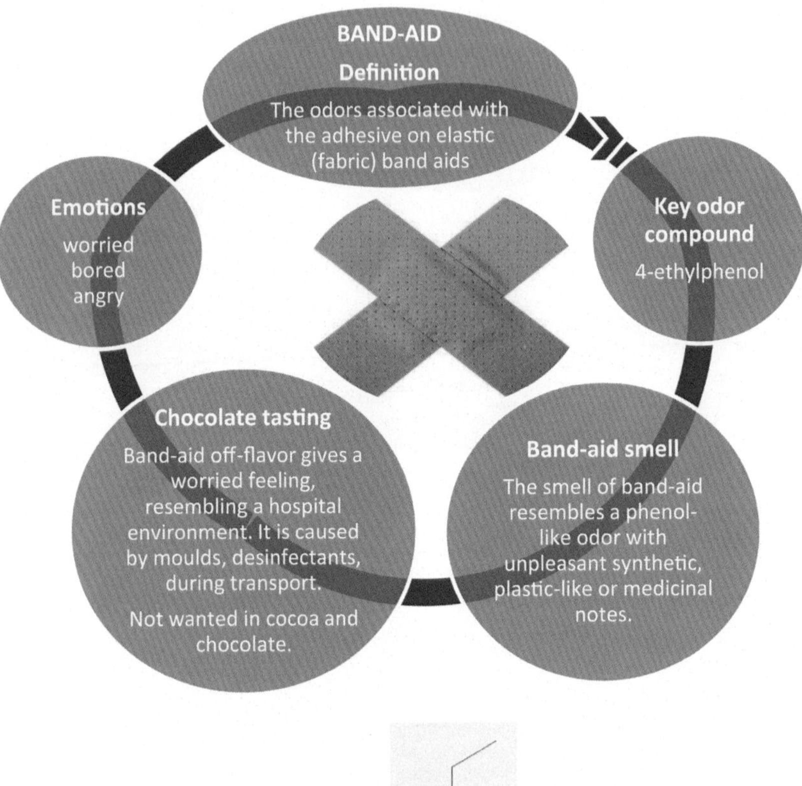

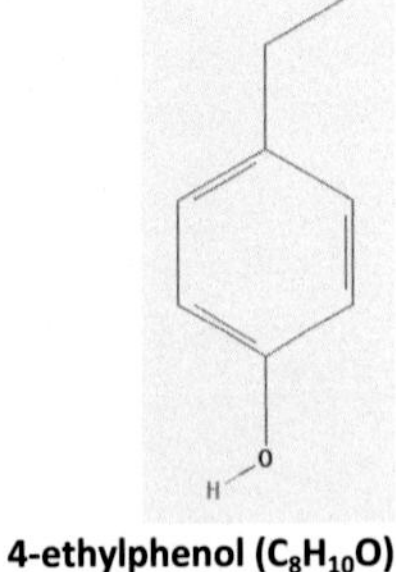

4-ethylphenol ($C_8H_{10}O$)

Soapy

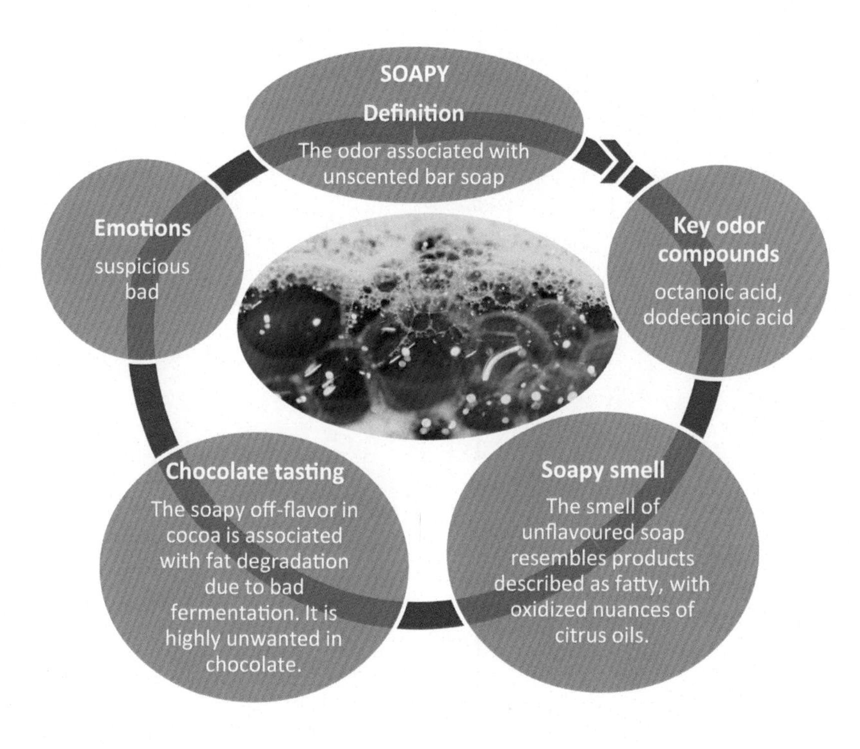

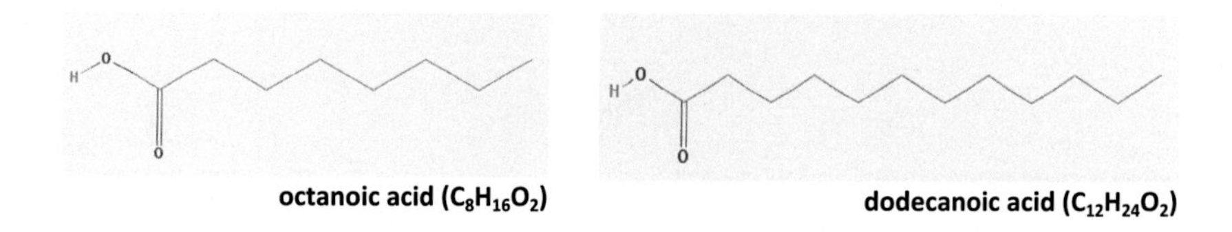

octanoic acid ($C_8H_{16}O_2$) **dodecanoic acid ($C_{12}H_{24}O_2$)**

Fishy

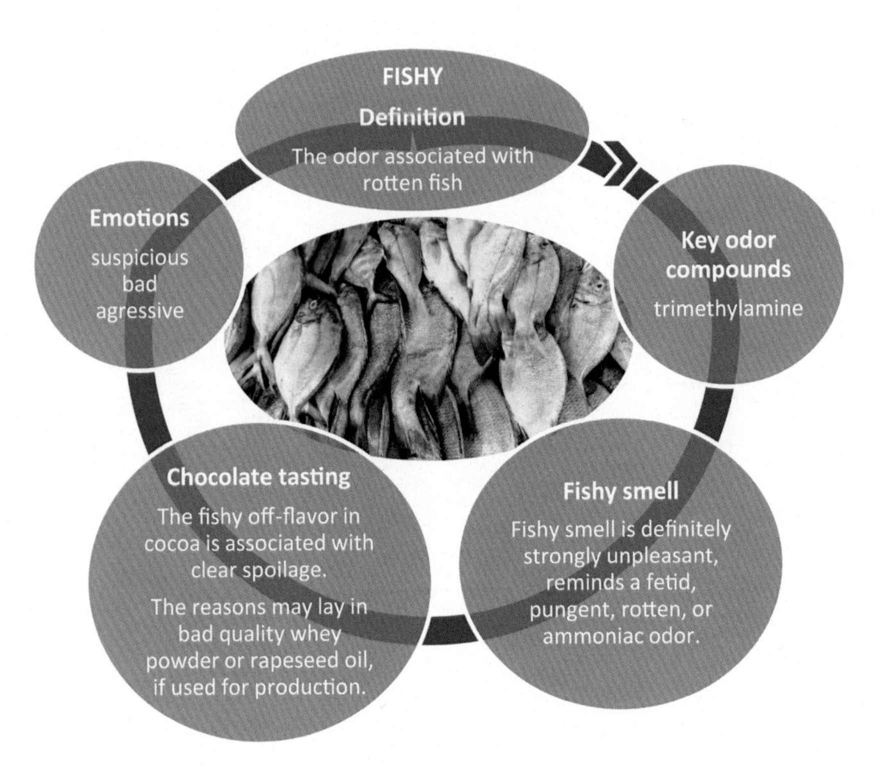

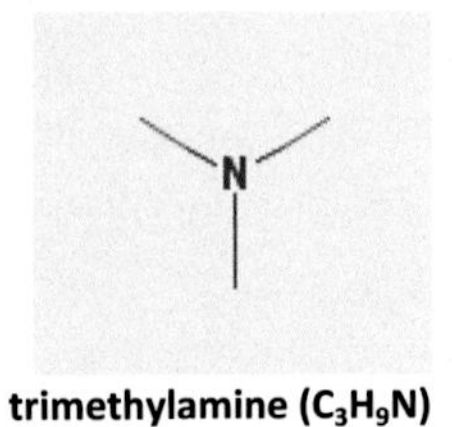

trimethylamine (C_3H_9N)

FOOD CONTACT MATERIALS (FCM)

Plastic

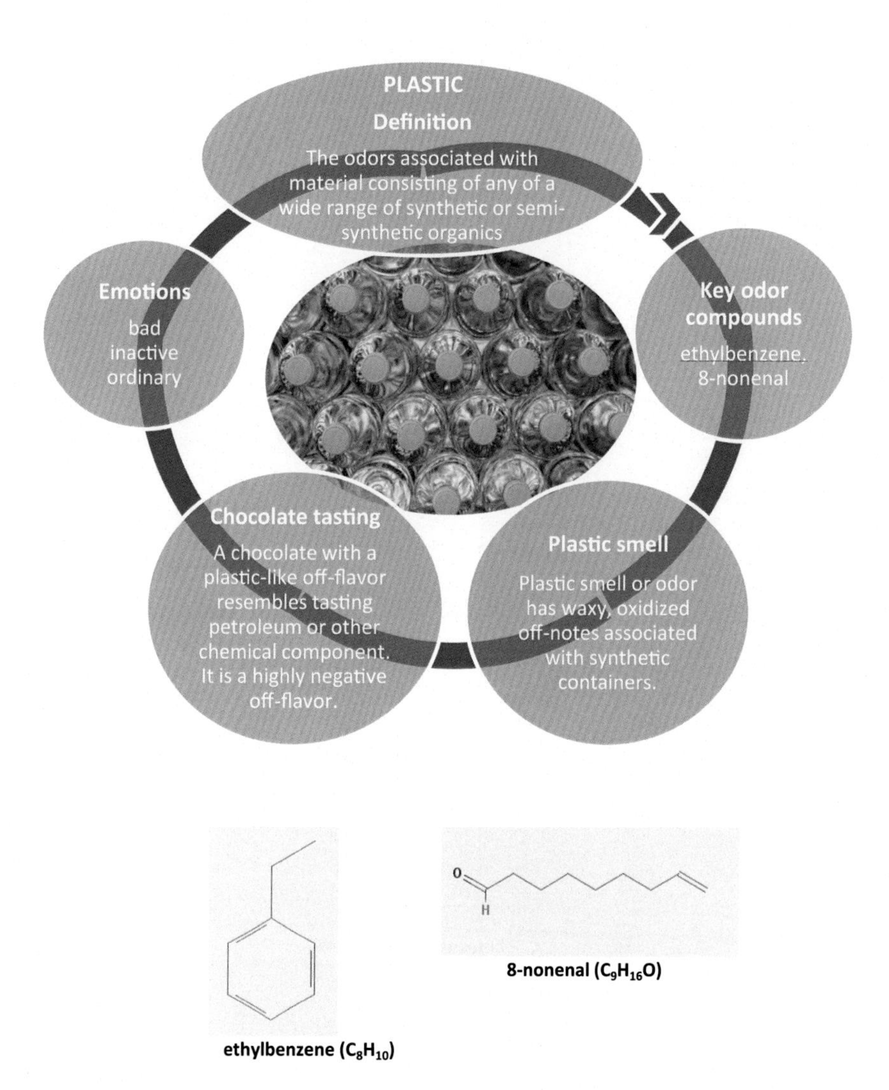

8-nonenal ($C_9H_{16}O$)

ethylbenzene (C_8H_{10})

Rubber

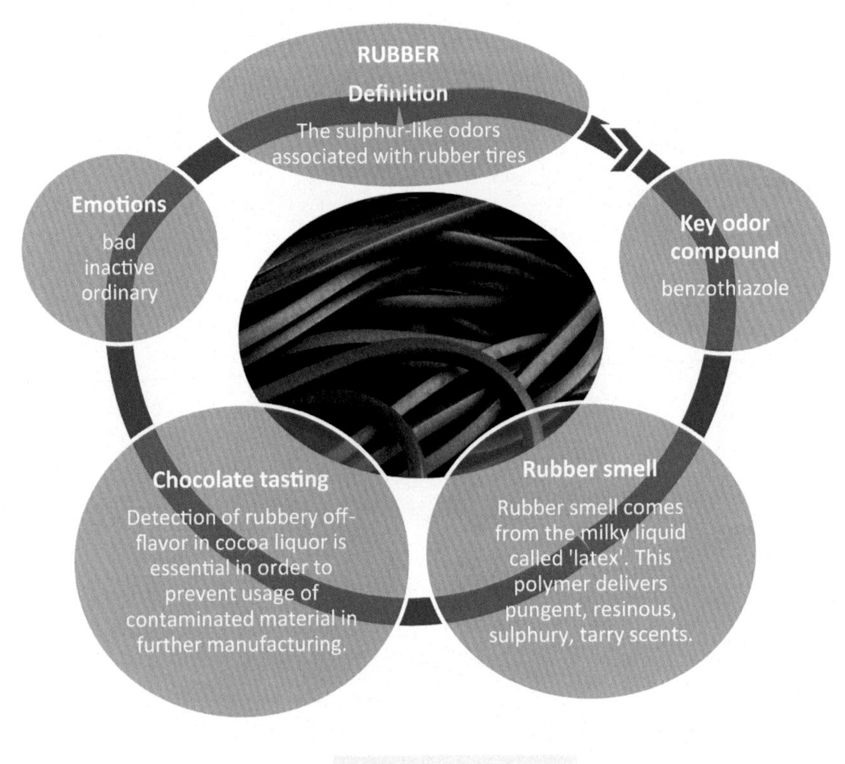

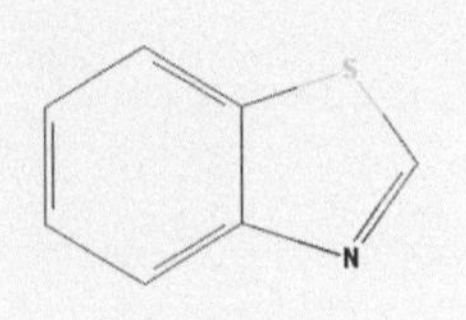

benzothiazole (C_7H_5NS)

Paint

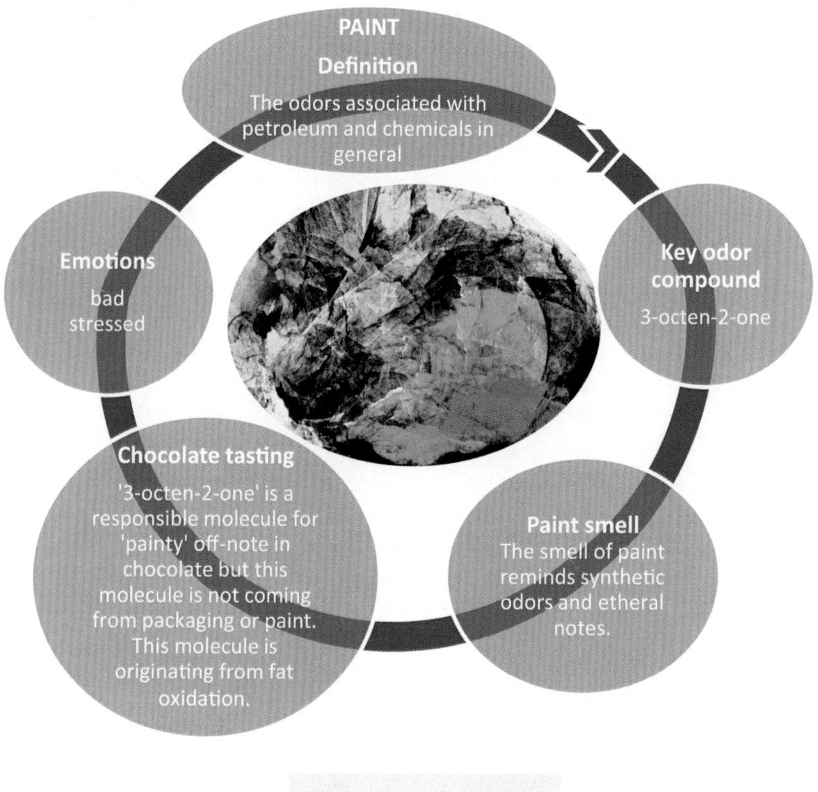

3-octen-2-one ($C_8H_{14}O$)

Cardboard

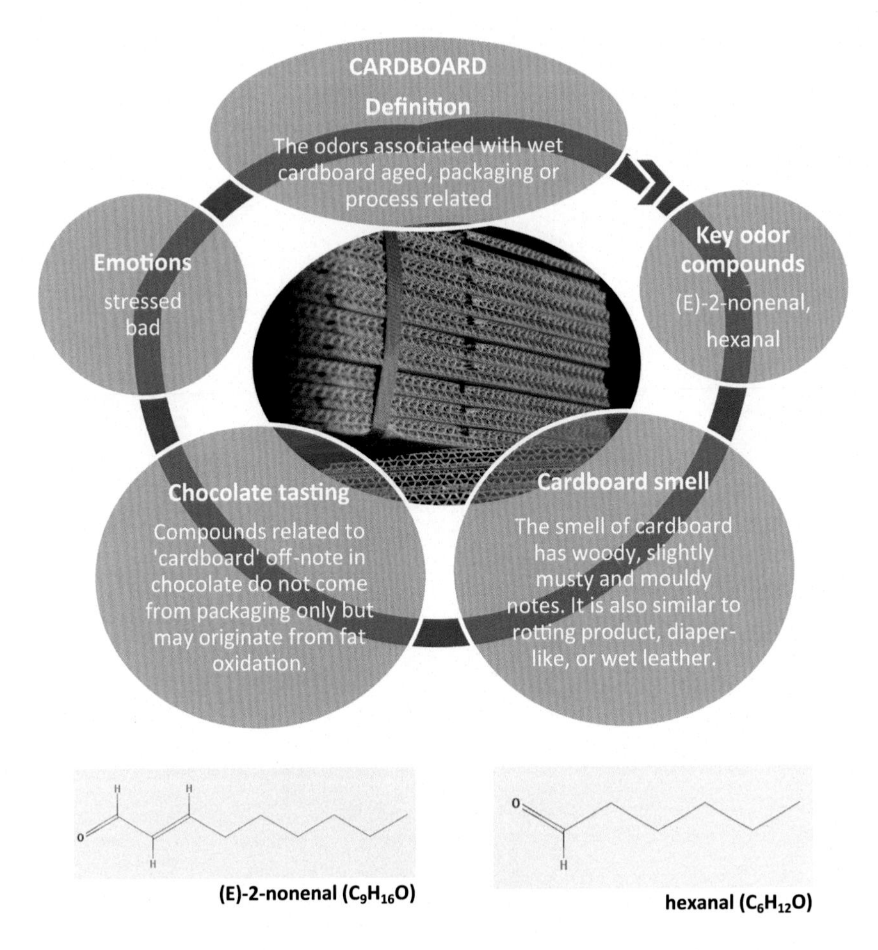

(E)-2-nonenal ($C_9H_{16}O$)

hexanal ($C_6H_{12}O$)

Metallic

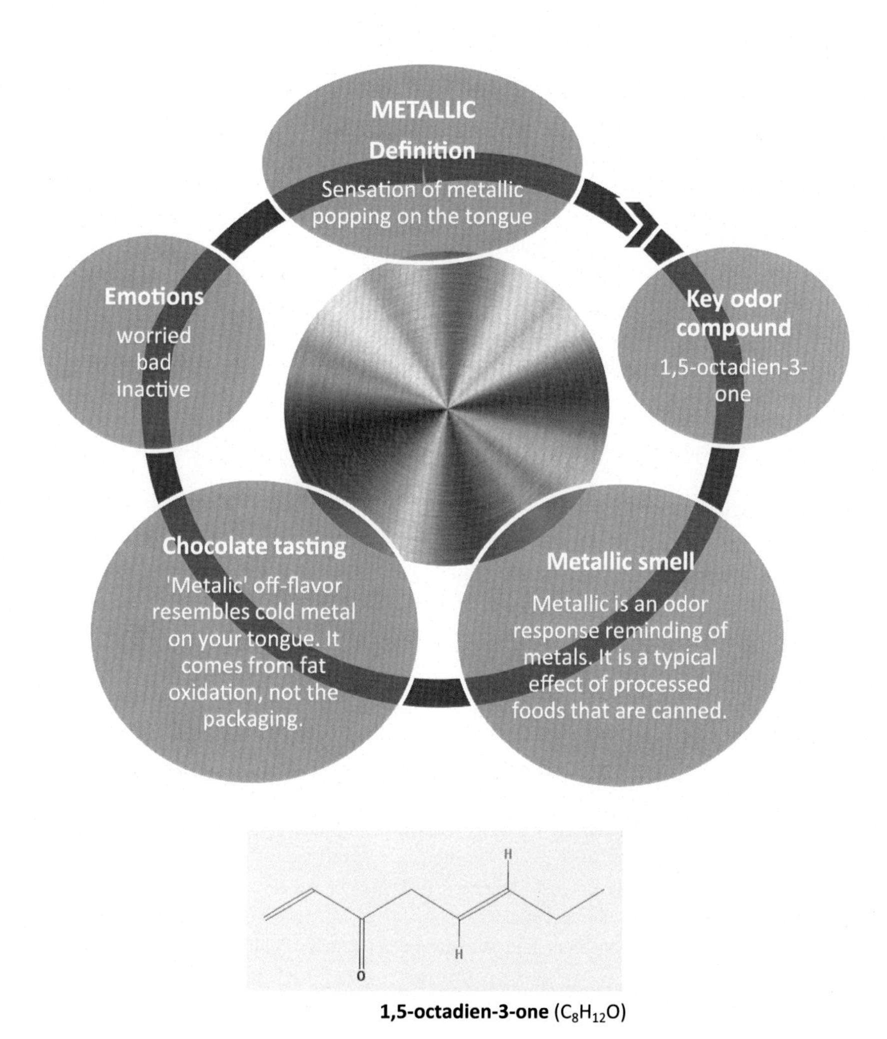

1,5-octadien-3-one ($C_8H_{12}O$)

FURTHER READING

Andrew, G.H., Piggott, L.J., 2013. Fermented Beverage Production. Springer Science & Business Media, Dordrecht, (barnyard).

Australian Food and Grocery Council (2007). Organohalogen Taints in Foods (medicinal).

Belitz, H.D., Grosch, W., Schieberle, P., 2009. Food Chemistry. Springer Science & Business Media, Germany, (earthy, straw/green, heated fat, metallic, rubber).

Brand, J.M., Galask, R.P., 1986. Trimethylamine: the substance mainly responsible for the fishy odor often associated with bacterial vaginosis. Obstet. Gynecol. 68 (5), 587–732, (fishy).

Carrau, F., Medina, K., Farina, L., Boido, E., Henschke, P., Dellacassa, E., 2008. Production of fermentation aroma compounds by *Saccharomyces cerevisiae* wine yeasts: effects of yeast assimilable nitrogen on two model strains. Fems Yeast Res. 8 (7), 1196–1207, (yeasty).

Charalambous, G., 1980. The Analysis and Control of Less Desirable Flavors in Foods and Beverages. Academic Press Inc., New York, NY, (soapy).

Czerny, M., Buettner, A., 2009. Odor-active compounds in cardboard. J. Agric. Food Chem. 57 (21), 9979–9984, (cardboard).

Davies, N., 2013. Environmental Analysis Monitoring for 'Taste and Odor Compounds' Including Geosmin and mib in Potable Water using the Agilent 7000 Triple Quadrupole GC/MS. Agilent Technologies Inc., USA, (musty/mouldy).

Decker, E.A., Elias, R.J., McClements, D.J., 2010. Oxidation in Foods and Beverages and Antioxidant Applications: Management in different industry sectors. Woodhead, Cambridge, (cardboard).

Diagnostic Key Postfermentation Classes of Off-Character Compounds. Viticulture & Enology (medicinal).

Frauendorfer, F., Schieberle, P.J., 2006. Identification of the key aroma compounds in cocoa powder based on molecular sensory correlations. J. Agric. Food Chem. 54 (15), 5521–5529, (oxidized, rancid).

http://www.thegoodscentscompany.com/ (hammy).

Jelen, H., 2011. Food Flavors: Chemical. Sensory and Technological Properties. Taylor & Francis, London, (fermented/alcohol).

Kim-Kang, H., 1990. Volatiles in packaging materials. Crit. Rev. Food Sci. Nutr. 29 (4), 255–271, (cardboard, plastic).

Park, Y.W., Haenlein, G.F.W., 2013. Milk and Dairy Products in Human Nutrition: Production. Composition and Health. John Wiley & Sons, Ltd, Oxford, (paint).

Reineccius, G., 2009. Off–flavors in foods. Crit. Rev. Food Sci. Nutr. 29 (6), 381–402, (burlap/jute bag).

Ridgway, K., Lalljie, S.P.D., Smith, R.M., 2010. Analysis of food taints and off-flavors: a review. Food Addit. Contamin. A 27 (2), 146–168, (paint).

Rowe, D.J., 1998. Aroma Chemicals for Savory Flavors. Perfumer & Flavorist, Oxford, (burnt, smoky).

Spanier, A.M., 2001. Food Flavors and Chemistry: Advances of the New Millennium. The Royal Society of Chemistry, Cambridge, (cheesy).

Wypich, G., 2013. Handbook of Odors in Plastic Materials. ChemTec Publishing, Toronto, (plastic).

Final Words

In 1957 the book 'The Hidden Persuaders' of Vence Packard was published. That work is still considered as one of the most influential books about advertising. More than 60 years later, we bring 'Hidden Persuaders in Cocoa and Chocolate'. We hope to convince the reader that some things around us are not always perceived consciously but often subconsciously. Most of our perceptions are linked to feelings, emotions, judgements and expectations.

Packard's 'hidden persuaders' are the messages, cleverly incorporated in advertising tools. Our hidden persuaders are precisely defined sensory attributes of cocoa and chocolate. They are defining the 'conscious wishes' or 'unconscious desires' of clients and customers in the chocolate business.

We recognize the need to switch from a 'maker-minded' to a 'market-minded' global industry that is built on strong sustainability and traceability systems, in which all stakeholders play respected roles. From cocoa farmers, through cocoa and chocolate production experts to end product developers, all of us need to understand each other's sensory language.

This book creates an opportunity for aligned communication and further attempts to unlock hidden messages in cocoa and chocolate. And it is not only about our professional audience, who already understands these hundred sensory descriptors. It is also about non-professional enthusiasts of cocoa and chocolate who hopefully are becoming more knowledgeable in order to taste more consciously and communicate their dreams and desires in a more effective way.

In the long run, this book can contribute to the fact that cocoa and chocolate products will resonate in the right market segments because the product developers, creative chefs and marketeers will better understand the customers and consumers' motivations.

This book's ambition is also to help switching from an often 'subconscious/emotional' to a more 'conscious/analytical' approach to the complex – but truly magical – *world of Cocoa and Chocolate.*

Finally, please remember that you may describe products with other words…

Amazing, ***Appealing,*** Appetizing, ***Awesome,*** Delectable, *Delicious,* **Delightful, Divine, Enjoyable, Enticing,** Excellent, **Exquisite,** *Extraordinary,* **Fantastic,** Finger licking, Gorgeous, **Great, Heavenly,** Lip smacking, **Luscious, Magic,** Marvellous, **Mouth-watering, Palatable,** *Pleasant,* Pleasing, **Satisfying,** Scrumptious, Superb, *Surprising,* Tantalizing, **Tasty,** *Terrific, Wonderful,* **Yummy...**

MOST OF ALL – PLEASE – ENJOY – COCOA AND CHOCOLATE!

Selected Reviews

Dr. Howard Moskowitz, President of Moskowitz Jacobs Inc., NY. USA

"Taste, smell, flavor – you can't be successful in the world of food unless you master them. Mastering them means understanding the science. And those who teach know that you get the most understanding when learning is a pleasure. Read this book, ENJOY learning about the science of taste, smell, flavor, as it is applied by the masters of chocolate, Barry Callebaut. When I founded the journal Chemical Senses in 1972 we had no joyful, elegant books like these. Today, more than 4 decades later, we see elegance mixed with science, in what I might (forgive me please) call 'a Delicious Mixture.' Bravo to the team at Barry Callebaut."

Dr. Boris Gadzov, FlavorActiV Ltd., the UK

"This book provides an excellent overview about Cocoa and Chocolate Sensory. It is very easy to follow the content and quickly find all the details related to the specific descriptor. The dashboard is synced with key info and icons for each sensory descriptor; this provides a very innovative all around presentation. The illustrations will aid any reader in any geography to quickly understand the meaning and importance of the sensory descriptors. I would like to thank the authors and recommend this book to any Cocoa, Chocolate producers and Sensory enthusiasts' readers, Students and Scientists."

Prof. Dr. Dorota Majchrzak, Department of Nutritional Sciences, University of Vienna, Austria

"Dear authors, I'm very pleased to congratulate you to this book that presents satisfactory and sufficient information about the sensory description of chocolate on different levels, e.g. include taste and aroma attributes as well as the trigeminal aspects or different emotions of consumers. The sensory descriptors comprise all possible flavor notes of cocoa beans, fruity, spicy and floral, which are positively associated with chocolate and takes also into account negative attributes, which have its origin in food contact materials. The colors of many very nice pictures, which are the additional explanation to the text, include e.g. the key aroma compounds, supported the text and are the strength of the book. Very important seems to be the general information about the basic tastes, with the specification of the corresponding receptors. Definition of flavor as a combined perception of taste and retronasal perception completed the content of the introduction. The book is suitable for application in both the industry as well as in research projects. Additionally, it can be very helpful for the consumers who are interested in the full range of the chocolate descriptors."

Index

Printed in the United States
By Bookmasters